A Practical Approach to
WBEM/CIM
Management

T0179155

A Practical Approach to
WBEM/CIM Management

Chris Hobbs

CRC Press
Taylor & Francis Group
Boca Raton London New York

CRC Press is an imprint of the
Taylor & Francis Group, an **informa** business

CRC Press
Taylor & Francis Group
6000 Broken Sound Parkway NW, Suite 300
Boca Raton, FL 33487-2742

First issued in paperback 2019

ISBN-13: 978-0-8493-2306-5 (hbk)
ISBN-13: 978-0-367-39454-7 (pbk)

Library of Congress Card Number 2003065978

Library of Congress Cataloging-in-Publication Data

Hobbs, Chris.
    A practical approach to WBEM/CIM management / Chris Hobbs.
        p. cm.
    Includes index.
    ISBN 0-8493-2306-1(alk. paper)
        1. Business enterprises—Computer networks. 2. Computer networks—Management. I. Title.

    HD30.37.H623 2004
    658′.054678—dc22   2003065978

Visit the Taylor & Francis Web site at
http://www.taylorandfrancis.com

and the CRC Press Web site at
http://www.crcpress.com

Dedication

In Memory of My Father
Ronald Stanley Hobbs
1922-2001
who taught me more than he knew.

Contents

SECTION III: INTERFACES

SECTION IV: PRACTICE

SECTION V: APPENDICES

Preface

Web-Based Enterprise Management (WBEM) and its Component Information Model (CIM) are coming of age. They appeared as a management tool for desktop computers in the late 1990s and have reached maturity at precisely the right moment: as the Simple Network Management Protocol—SNMP—has become inadequate to meet the rising demand for device and service management at higher levels of abstraction.

I intend this book to be used by those, perhaps with experience with the SNMP or the Telecommunications Management Network (TMN) protocols, who need to become familiar with WBEM/CIM.

I have included some high-level material, particularly in the early chapters, outlining some advantages of WBEM/CIM which would be of interest to managers or people working in marketing organisations, but my primary target readership is the working system architect and engineer who will be designing WBEM-based systems and writing the code for a management application.

Acknowledgements

I gratefully acknowledge the help given to me by many engineers in producing this book—in particular Colin Ashford, John Bell, Sharon Chisholm, Tom Chmara, Eugene Deery, Randy Mortensen, Francis Ovenden, and Ying Zeng. They have contributed much during whiteboard discussions. I have enjoyed working with Ying on several CIM implementations, learning together as we examined code and took our first, faltering steps into provider writing. Francis, Randy, and Colin have provided direct constructive criticism of the text, ranging from correction of factual errors to broad hints about possible changes in style. Francis must have a special mention—his meticulous reading of the manuscript pointed out technical oversights, unclear explanations and many, many typos. I am particularly grateful to him for this work.

I would also like to thank my wife, Alison, for her help in improving the readability of the book by the reduction (but, in spite of her efforts, not the elimination) of jargon and the simplification of sentences. Her interest in WBEM and CIM is perhaps not as strong as mine and she has sometimes found wakefulness challenging during proof reading. On one version of the manuscript in a Frequently Asked Questions section, I found a penciled suggestion in her handwriting which proposed the additional question, "Why am I reading this?"

I have received numerous insights (and sources of Frequently Asked Questions) from members of the WBEM/CIM community who have unwittingly contributed to this book during face-to-face and e-mail discussions.

In various sections of the book, in particular in Chapter 6, I include examples taken from the Core and Common Models standardised by the Distributed Management Task Force (DMTF). This material is reproduced with the permission of the DMTF. A full description of the licence under which this code may be reused is given in Appendix H.

There are two further teams that have unconsciously helped enormously in the preparation of this book: Donald E. Knuth and the team that created LaTeX 2_ε and the team that produced the *dia* drawing tool. I prepared the book using LaTeX 2_ε and the line drawings using *dia*. The ease of using these tools and the simplicity with which they interworked really made the book possible.

Chris Hobbs
Ottawa

The Author

Chris Hobbs has worked in telecommunications since 1973. He moved from the circuit switching of 10-character-per-second paper tape readers to packet and cell switching in X.25, Frame Relay, ATM, and IP networks. In the transmission field he has worked through the evolution of asynchronous, plesiochronous, and synchronous networks to today's switched photonic networks.

Most of the technical problems which caused the shift from circuit to packet switching having been resolved, Chris is looking forward to the return of circuit switching and hopes to be present on the day (say, 24th October 2014) when the last packet travels over the Internet. This will be a sad day for many in the IETF, but will be a day of wonder for Internet users.

All of these networks, the fanciful, packet-free Internet included, require management. Perhaps more important, the services running on those networks require management. Over the years, Chris has worked with many *ad hoc* management systems as well as those standardised by the ITU-T and IETF.

Telecommunications has taken Chris (and his adventurous family) to work in England, Holland, Switzerland, Wales, the United States of America and now Canada. He has worked for national PTTs, consultancy companies, and telecommunications equipment manufacturers, and is currently employed by Nortel Networks in Ottawa.

The work being done on WBEM/CIM within the DMTF and elsewhere has excited Chris and appears to him to be a tremendous opportunity for a whole industry. WBEM is an initiative which at last seems to have learned from the past—not from past mistakes, because CMIP and SNMP were not mistakes—but from past experience.

Having been hampered himself by the lack of a clear text describing the WBEM/CIM architecture at a working level, Chris hopes that this book will fill that gap.

Chris programs in C++, C, and Python and is building up his strength to tackle Java (the size of the books on that subject being daunting).

Outside of telecommunications, Chris is a commercial pilot and teaches students to fly—see his previous book on the pleasures of flying. He is also deeply interested in the interpretation of Schubert Lieder, particularly the *Winterreise* cycle.

MANAGEMENT

Chapter 1

Introduction

The Aim

I am a working engineer designing management systems for telecommunications products. This is the book that I needed when I first started to work with WBEM/CIM.

The Subject

WBEM/CIM is a hot topic in a number of areas, including the storage network, electrical power supply, desktop computing, and telecommunications industries. These industries are seeking a common and flexible way of managing a collection of heterogeneous devices and services. WBEM/CIM provides this standardisation and addresses the management of services in particular.

WBEM/CIM comprises a device and service management architecture, a language for describing the management features of a device or service, a set of interfaces for accessing the management information, and an extensive set of common models for storage, computing, and telecommunications applications.

The Readership

I intend this book to be of practical use to those, perhaps with an SNMP or TMN background, who need to become familiar with WBEM/CIM to design a WBEM management system, to produce a CIM model of a device or service, to implement a WBEM management system, or to interwork with one.

Some of the early chapters compare WBEM/CIM with SNMP and TMN, and this material also would be of interest to managers and people working in marketing organisations. My primary target readership, however, is the working system architect and engineer who will be designing CIM-based systems and writing CIM clients and providers.

I work in the telecommunications field, and this book inevitably has a telecommunications flavour, but it is not only aimed at a telecommunications readership. Where I have used a telecommunications example which might not be clear to all readers, I have tried to explain the basic concepts in the Glossary starting on page 307. The book is designed to be of value to anyone using WBEM and CIM, whatever their industry.

Similarly, I am a Python and C++ programmer. Coding examples are restricted to Chapter 12 and are only given in C++. I have, however, kept the code very simple and translation into Java for use with a Java-based WBEM server should not be difficult.

The Book

I have structured this book into four parts:

1. The first part describes the purpose of device and service management and compares different architectures.
2. The second part, starting with Chapter 4, describes in detail the WBEM/CIM architecture and the standard models developed by the DMTF.
3. The third part, starting with Chapter 7, describes the interfaces with which your WBEM/CIM code will have to work.
4. The fourth part, starting with Chapter 9, delves deeper and gives examples of defining your own models and writing your own code. It also gives examples of some of the tools which are available to help you.

In more detail, this book answers the following questions:

■ Chapter 2: What are device and service management? What problems do they set out to solve?
■ Chapter 3: What is WBEM/CIM? What is the distinction between WBEM and CIM? Why do we need a new standard when we have SNMP and TMN? How does XML fit into all of this?

- Chapter 4: What are the components of the WBEM architecture? How do they fit together at the highest level?
- Chapter 5: What is a model and how is it defined in CIM? What standard models are available?
- Chapter 6: What standard models does the DMTF provide? What is the process for introducing new models?
- Chapter 7: How does a WBEM client talk to its WBEM server?
- Chapter 8: How does a WBEM server deliver events and alarms to an external program?
- Chapter 9: How does one set about building a model for a device or service?
- Chapter 10: What pitfalls has the author fallen into while modelling?
- Chapters 11 to 13: How are the information and processes unique to a particular problem encoded and merged with the standard WBEM/CIM components? How are the providers and clients coded and linked into the WBEM server?
- Chapter 14: When starting from an existing SNMP (or proprietary) management system, is it possible to transition gently to the WBEM/CIM approach?
- Chapter 15: What tools exist to help with the design and implementation?
- Chapter 16: Is it better to buy or build the software necessary for a WBEM system? What options are there for obtaining this software?

At the end of most chapters, I have included a section called *Frequently Asked Questions* (FAQs) to answer some of the specific questions that might arise from reading the chapter. Most of these questions have been asked during presentations I have given or recur in the mailing lists associated with WBEM/CIM implementations. I have also used the FAQs as a repository for isolated facts related to the topic of the chapter which do not fit easily into the main text.

The Glossary, Appendix G, contains definitions of some of the terms which I have used which may not be familiar to all readers.

The Moving Target

If this book is to be a practical text, then it must aim at a moving target—new CIM models are appearing regularly from the DMTF, new companies are emerging almost hourly to supply tools and software to the new industry, and the open source implementations of the architecture are being enhanced regularly. I have tried to avoid focussing on details prone to rapid change, but a book that never descends to implementation detail is not useful. Where I have described interfaces or models not yet standardised, I have tried to point this out.

In particular, examples from the DMTF's models (schemata) in this book use the preliminary issue of version 2.8.

WBEM Implementations

There are numerous commercial and open source implementations of WBEM/CIM—11 are listed in the WBEMSource software directory at http://www.wbemsource.org/registry/.

In most chapters, I have tried to avoid using information specific to any particular implementation. This has not been possible in Chapter 12 where I give coding examples. Having to choose a particular implementation, I selected the one with which I have the most experience: openPegasus. This implementation is open and the code is freely available to everyone from the openPegasus Web site (http://www.openpegasus.org). Other C++ implementations require very similar code.

The DMTF

The organising body for WBEM and CIM is the Distributed Management Task Force—DMTF—an industry body formed to lead the development, adoption, and interoperability of management standards. Currently the DMTF membership includes over 100 companies and approximately 50 academic institutions.

All of the models and standards which I quote in this book are freely available from the DMTF's Web site (http://www.dmtf.org/home).

Chapter 2

Device and Service Management

Device and Service Management

The topic I address in this book already has many acronyms: CMIP, SNMP, ASN.1, NMS, EMS, OSS, OSF, TMN, OSI, MIB, FCAPS, GDMO, and SAP. Before I add CIM, WBEM, CIM-XML, xmlCIM, UML, *mof*, NPI, CMPI, SBLIM and other ingredients to the alphabet soup, I would like us to agree on the topic we are addressing and some basic, simple vocabulary to describe it.

Imagine that I have invented a toaster which can accept simple management commands such as "lower the bread, turn on the heaters, turn off the heaters, tell me the colour of the bread (never mind how it does this), eject the bread" and "set thermostat to $x°C$."

Toaster management in its simplest form is illustrated in Figure 2.1: I sit at a management workstation (my computer) which is locally

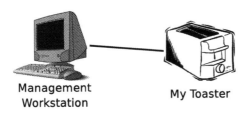

Management
Workstation

My Toaster

Figure 2.1 Basic Toaster Management

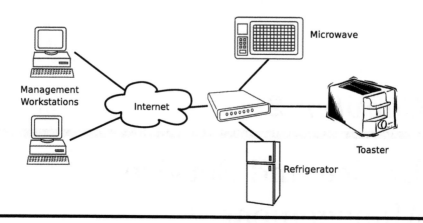

Figure 2.2   Kitchen Management

connected to the toaster and enter commands to cook my toast. The protocol that my management workstation uses to talk to the toaster is a private matter; the interface presented to me is tailored specifically to the toaster and, because the connection between the two is local, any threats of unauthorised people using my toaster may be handled by locking my kitchen door.

These days, however, all devices must be connected to the Internet; so, after a period of private toast making, I succumb to peer pressure and connect my toaster and other kitchen appliances to the wider world.

Figure 2.2 illustrates my new situation: I can now manage my toaster, microwave, and refrigerator from management workstations in my office or local Internet café. However, I now have more problems to face: the remote connection means that locking the front door is no longer sufficient to prevent unauthorised toast making; the heterogeneous nature of the devices means that some form of standard protocol is needed to keep the management workstation software tractable, and the opportunity for my wife and me both to be managing the kitchen at the same time means that we may conflict with each other when configuring the same device.

In addition, because there are several devices involved, we may want to manage not only the individual devices, but also services offered across several of them: co-ordinating, for example, microwaved beans with golden toast. How may such services be managed now that they are not resident on any single device?

These are the kind of problems, although not necessarily in the kitchen, which the DMTF set out to address in creating the WBEM/CIM architecture.

Given the confusion of terminology within the device, network, and service management world, I intend to use only the terms "device management" and "service management" in this book.

Device management is the term I use to mean the management of a single device using commands which the device itself knows how to execute. The toaster command "lower the bread" is a device management command as it refers to a primitive action which the toaster can perform. Note that, as an operator, I do not know how the command is executed by the toaster; it may cause a voltage to be placed on a relay coil, which causes a rod to move outwards, causing a micro-switch to close, and so on. It is sufficient for me to know that lowering the bread is a primitive operation that the toaster can do.

Service management is the term I use to mean the management at a higher level of abstraction. The command "toast a piece of bread the way Chris likes it and inform me when it is done" is an example of service management. This is not a command that the toaster knows how to handle directly—some intermediary process will have to break the command into primitive steps (lower the bread, turn heater to 150°C, etc.) before the toaster can do it. In this case the service management command affects only one device, but typically such a command might affect several: "cook a meal of beans and toast the way Alison likes it."

Of course, this categorisation is not precise. One person's service management is the next person's device management; a telecommunications carrier might use service management commands to configure a virtual IP router. The customer of the virtual router would then see it as a device and perform device management on it.

The reason I have stressed the device/service distinction is that, in many fields, customers are demanding more service-level management.

In the telecommunications and computer worlds, for example, buyers are putting increased pressure on suppliers to provide service management using abstractions that are meaningful to the buyers' operators. They are demanding that they be able to manage video streaming applications, mobile workers, and other services affecting many parts of a network so that operators have to work less often at the device level; e.g., "set parameter 7 of port 39 on device 87 to 1."

Operators capable of manually converting requests for a video streaming application into a set of primitive commands to configure parameters in devices are rare and expensive as they need to be experienced and have a deep insight into the system. The cost of these operators, the so-called OPEX or operational expenditure, constitutes a large part of a network owner's outlay, and some telecommunications operators have recently announced that, with the next generation of network equipment, they are looking for a reduction of between 25 and 40 percent in this cost. Even with experienced operators, however, manual configuration using low-level commands is error prone. Although we still need to be able to configure the parameters of a particular port (device management), the cost of skilled operators and the revenue lost through manual configuration errors have driven a demand for the management of end-to-end services and abstractions.

This move to high-level services is reflected in the work of organisations such as the Parlay Group* and JAIN,† which were formed to allow network operators to define open application program interfaces (APIs) to networks, particularly Internet Protocol (IP) and Public Switched Telephone Networks (PSTNs).

It is important to re-iterate that one organisation's service is another organisation's device; the services being discussed by Parlay and JAIN, for example, would not be considered services by a customer of a company providing a distributed Backup and Restore service—they are *components* which allow the Backup and Restore service to be built.

The cost of operator error, particularly when operators are asked to work at the component level, has also initiated a discernible move away from operators entering commands ("disable port 7 on card 23") to entering suggestions ("I suggest that port 7 on card 23 be disabled"). The management system can then reply that disabling that port would affect the service of 453 users, currently producing a revenue of $1087 per minute. The management system could then refuse to obey the command for anyone other than a senior operator.

Because many system failures are caused by inappropriate operator commands (an oft-quoted Gartner study‡ attributes 40 percent of all failures directly to operator error and a further 40 percent to software bugs, many stimulated by management tasks such as upgrading), the softer "suggestion" mode coupled with a higher-level abstraction allows management software to protect operators from their own errors.

In addition to the device/service split, there are many ways of categorising types of management; Figure 2.3 illustrates some of these (although there are many more). The axes shown are:

---

* http://www.parlay.org/
† http://java.sun.com/products/jain/
‡ Gartner Research Note TG-07-4043

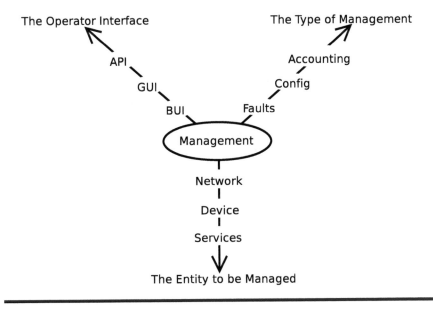

Figure 2.3   Some Management Axes

- **What type of management** do you actually want to do? For example, do you simply want to collect and display alarms or do you also want to configure things? Do you want to collect and analyse performance information? This classification of management is often known as FCAPS (fault, configuration, accounting, performance and security management). I describe FCAPS in more detail in Appendix C.
- **The entity to be managed** might just be a single device, but also might be multiple devices one at a time, or multiple devices simultaneously to establish a connection across a network. There are numerous ways of enumerating these scopes; the Telecommunications Management Network Forum, for example, uses the categories shown in Figure 2.4.[§]
- **Through what type of interface** do you want to manage your system? By using a graphical user interface (GUI) with sophisticated diagrams, pull-down menus, etc.? By typing textual commands into a command-line interface (CLI)? By writing a

---

[§] Note that this picture is often drawn as a pyramid the other way up with business management very small at the top. I have changed the representation because it seems to imply that element management is large, network management slightly smaller, etc. In fact problems increase rather than decrease as one travels towards business management.

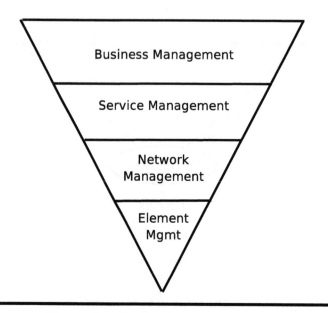

Figure 2.4   TMN Management Layers

program to access an API? By using a standard Web browser (browser user interface: BUI)? Each of these may be applicable in different contexts and may need to be used by different operators managing the same device simultaneously.

Many devices, particularly small routers designed for homes and small offices, are being shipped with BUIs: the operator simply connects an HTML browser to a particular port on the device and enters information into the screens which appear. In this case, the entire management system is driven by software running on the device being managed. More sophisticated systems may require a Java or C++ program on the operator's computer to provide a more complete graphical interface for the operator—showing, for example, pictures of devices turning red when faults are detected and allowing the operator to point and click on components to be managed.

Even more sophisticated systems may require a CLI for the operator and a programmatic interface for higher-level management systems. There is certainly much anecdotal evidence to suggest that, at least within the telecommunications field, "real" operators do not use graphical interfaces and that this is not just a macho attitude—it is a reflexion of the speed and flexibility of a command line for an experienced operator.

Frequently Asked Questions

FAQ 1 *Is network management the same as network control?*

Gulliver's Travels can be interpreted at several levels: it is both a fascinating book for children and a biting political satire ("my principal Design was to Inform, and not to amuse thee," said Jonathan Swift).¶ Similarly there at least two levels of answer about the distinction between management and control of devices or networks. First there is a clear distinction:

■ Management is carried out by an operator sitting at a management station. He or she enters commands to reconfigure equipment and those commands are carried out.

■ Control, by contrast, is carried out autonomously by the network, without operator intervention in response to some external stimulus. For example, a link fails and the network automatically reconfigures equipment to route around it. I pick up my telephone and dial a number in Vancouver. The network detects this without operator intervention and reconfigures equipment to route my call.

There is, however, a deeper level at which these operations are similar, if not identical. Both can be thought of as a stimulus (which could be an operator command) causing changes to the device.

FAQ 2 *Why the toaster?*

There is a classical joke comparing "real" engineers with object-oriented computer architects with which you are probably familiar. In case you are not, with acknowledgements to the anonymous author, here goes.

Once upon a time a king decided to test his two chief advisors. He showed them a toaster and asked them to design an embedded computer for it.

The first advisor, an engineer, replied, "Using a four-bit microcontroller, I would write a program that reads the darkness knob and quantises its position from snow white to coal black. The program would use that level as the index to a table of timer values. It would

---

¶ It is also the source of the Endianness problems which have plagued computing, but that piece of programming trivia is irrelevant here.

turn on the heating elements and start the timer with the initial value selected from the table. At the end of the time delay, it would turn off the heat and pop up the toast. Come back next week, and I'll show you a working prototype."

The second advisor, a computer specialist with object-oriented design training, immediately recognised the danger of such short-sighted thinking. He said, "Toasters don't just turn bread into toast; they are also used to warm frozen waffles. What you see before you is really a breakfast food cooker. As your subjects become more sophisticated, they will demand more functionality. They will need a breakfast food cooker that can also cook sausages, fry bacon, and scramble eggs. A toaster that only makes toast will soon be obsolete. If we don't look to the future, we will have to redesign the toaster completely in just a few years.

"With this in mind, let's formulate a more intelligent solution to the problem. First, create a class of breakfast foods. Specialise this class into subclasses: grains, pork, and poultry. The specialisation process should be repeated with grains divided into toast, muffins, pancakes and waffles; pork divided into sausage and bacon; and poultry divided into scrambled eggs, hard-boiled eggs, poached eggs, fried eggs and various omelette classes.

"The ham and cheese omelette class is worth special attention because it must inherit characteristics from the pork, dairy, and poultry classes. Thus, we see that the problem cannot be properly solved without multiple inheritance. At run time, the program must create the proper object and send a message to it that says, 'Cook yourself.' The semantics of this message depend, of course, on the kind of object, so they have a different meaning to a piece of toast than to scrambled eggs.

"Reviewing the process so far, we see that the analysis phase has revealed that the primary requirement is to cook any kind of breakfast food. In the design phase, we have discovered some derived requirements. Specifically, we need an object-oriented language with multiple inheritance. Of course, users don't want the eggs to get cold while the bacon is frying, so concurrent processing is required, too.

"Don't forget the user interface. The lever that lowers the food lacks versatility, and the darkness knob is confusing. Users won't buy the product unless it has a user-friendly graphical interface. When the breakfast cooker is plugged in, users should see the company logo appear on the screen, and when they click on it, the message 'Booting Linux 2.8' appears (Linux 2.8 should be available by the time the product gets to the market). Users can use a menu to select the foods they want to cook.

"Having made the wise decision of specifying the software first in the design phase, all that remains is to pick an adequate hardware platform for the implementation phase. A 1.2-GHz processor with 128 MB of memory, a 30-GB hard disk and a colour VGA monitor should be sufficient. If you select a multitasking, object-oriented language that supports multiple inheritance and has a built-in GUI, writing the program will be a snap. (Imagine the difficulty we would have had if we had foolishly allowed a hardware-first design strategy to lock us into a four-bit microcontroller!)"

The king had the computer specialist beheaded, and they all lived happily ever after.

I hope that, as you travel through this book, savouring its object-oriented flavour, you will keep your head.

Chapter 3

WBEM and Other Management Systems

As you read this chapter, remember the two types of fool: the one who says, "because it's old, it's good" and the one who says, "because it's new, it's better."

WBEM and CIM

WBEM is perhaps the most misleading title since "Miss Universe." I believe that only terrestrial competitors have ever taken part in the Miss Universe competition and I spend a measurable part of my life explaining to colleagues that WBEM has nothing to do with the World Wide Web* and can be applied to the management of systems unrelated to enterprises (meaning companies). Much of my work is with carriers, the owners of long-distance telecommunications networks. Such networks have to meet stringent availability requirements and, within this world, "enterprise grade" is contrasted with "carrier grade," the enterprise epithet implying lower quality.

The M in WBEM, meaning management, may also be misleading; it refers to the management of devices and services rather than people, e.g., setting up voice mail for a particular subscriber, intercepting alarms from storage devices, tracing a fault in the address scheme of

---

* Except that it borrows one protocol: HTTP.

a Local Area Network, installing a new version of a word processing package on everyone's computer within an enterprise, and so on.

WBEM is the product of the DMTF and this group might therefore be the definitive source on what WBEM constitutes. In a slide presentation in June 2000 on the role of the DMTF, Winston Bumpus, the DMTF President, explicitly said this role was desktop systems, enterprise networks, and E-business infrastructure, and this is indeed the current emphasis.

Perhaps my main criticism of the DMTF is that it has set its sights too low. As I hope I demonstrate implicitly, the WBEM architecture, language, and models are very close to being ready for the higher-quality carrier domain. I discuss my only reservation in my answer to Frequently Asked Question 28 on page 149.

Having said what WBEM is not, I should now say what it is. WBEM encompasses the architecture and technologies for device, network, and service management. It emerged in the mid to late 1990s as a means of managing desktop computers, and in the late 1990s it started to evolve into a more general-purpose management tool.

The two major management standards in use in networks today are SNMP, primarily used for data (IP) networks, and TMN, which has traditionally been used in the backbone telecommunications networks. The rise of data networking and the convergence of data and voice networks mean that these two systems are now colliding in some networks, generally introducing SNMP management into the backbone rather than TMN into the enterprise. WBEM can be seen as a way of upgrading and ultimately replacing both.

The key differences between WBEM and the more traditional management standards such as TMN and SNMP are the available modelling constructs and WBEM's clear separation of the interface used to access information from the model of the device being managed.

SNMP has a very simple modelling language, allowing objects to be defined and interrelated through a "contains" hierarchy. Apart from its ability to construct two-dimensional tables, there is no other way to associate and combine items.

In contrast, both TMN and CIM have sophisticated object-oriented structures supporting concepts such as inheritance, classes, and polymorphism. Among other advantages, such as more intuitive modelling, these techniques allow much larger models to be built without them becoming unwieldy.

WBEM comprises a set of standardised technologies (see Figure 3.1) that include:

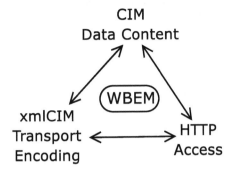

Figure 3.1   WBEM Components

- CIM: A data modelling process and language (known as Managed Object Format (*mof*)). CIM also includes a number of standard models (schemata) for systems, applications, networks, devices, and other common components, expressed in the *mof* language; see Table 3.1. This standardisation enables applications from different developers on different platforms to describe management data in a standard way so that it may be shared among a variety of management applications.
- A standardised architecture for hierarchical management systems including the definition of a WBEM server (see Figure 4.4) and its interfaces. I describe this architecture in detail in Chapter 4; suffice it to say here that the WBEM server acts as a broker between the operators' workstations and the complex systems under their management.
- xmlCIM, an encoding specification, for encoding commands and responses which can be used to represent WBEM entities:
  - The definition of XML elements in Document Type Definition (DTD).
  - The representation of CIM in XML Specification.
- HTTP access, the HTTP encapsulation (CIM-XML), the transport mechanisms for carrying commands and responses across a network, including the "CIM Operations over HTTP" and the anticipated "CIM over SOAP" (Simple Object Access Protocol, as standardised by the W3C, the World-Wide Web Consortium) specifications.

Model	Contents	Reference
Applications	Software products and applications and their deployment and monitoring	-
Database	A database and the database service (as in RFC1697)	-
Device	Configuration and state information for hardware	-
Event	Events and alarms	Page 151
Interop	Management of the WBEM server itself	Page 250
Metrics	Information measured in a system and the handling of it	-
Network	Networks and network protocols (OSPF, BGP, etc)	Page 100
Physical	Inventory and asset management	-
Policy	Rules and associated actions	Page 107
Support	Interaction of support models for different equipment	-
System	The combined action of several components working together as a system	Page 99
User	The location, identity, and authority of different users	Page 114

Table 3.1   CIM Common Models

The Need for New Management Standards

Today, in addition to a large number of *ad hoc* management systems, three comparable standardised architectures are in use in telecommunications, storage, and computing: SNMP, TMN, and WBEM. I describe these in outline in the sections which follow.

The need for standardisation, at least at the interface to the device being managed, has led to a number of proposals for fixing part of the problem. These have tended to exclude the model of what is being managed, and to deal only with the mechanics of the interface between the operator's workstation and the managed device. The Internet Engineering Task Force's (IETF's) NetConf initiative can be seen as a partial solution of this kind—it defines a vendor-independent protocol (XML over SOAP or BEEP—Blocks Extensible Exchange Protocol— over TCP) and a set of basic commands that can be sent to a managed device, but in fact it is just a standardised layer above the device vendor's command-line interface (CLI).

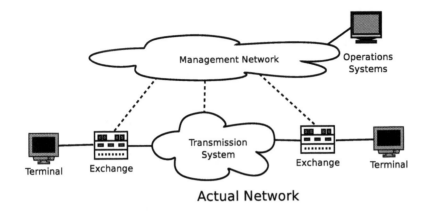

Figure 3.2 The Telecommunications Management Network

Telecommunications Management Network (TMN)

In 1985, the standardisation section of the International Telecommunication Union (ITU-T), then known as the CCITT, began to define a set of recommendations for managing telecommunications networks. These emerged in 1992 under the title M.3010, and an update was issued in 1996.

M.3010 starts, not with the model, but with the definition of a *management* network that overlays the telecommunications network being managed, connecting with it at different points—See Figure 3.2. As can be seen from the figure, the interface points between the TMN and the telecommunications network are formed by exchanges and transmission systems. For the purposes of management, these exchanges and transmission systems are connected to so-called operations systems, management systems to which operator workstations may be attached.

M.3010 defines the general TMN management concepts and introduces several management architectures at different levels of abstraction. Of these, two are of interest when comparing TMN with CIM and SNMP. TMN has:

■ An information architecture based on ITU-T Recommendation X.720, "Information Processing Systems—Open Systems Interconnection-Structure of Management Information – Part 1: Management Information Model," Geneva, 1993. The modelling semantics defined in this document are very similar to those of CIM, including the concepts of classes with properties (known as attributes) and behaviour. CIM can be considered an extension of this modelling methodology, incorporating more recent thinking on object-oriented specification. As the CIM modelling philosophy has its associated language, *mof*, so the X.700 series of specifications has its language, "Guidelines for Definition of Managed Objects" (GDMO), which is defined in X.722. As with CIM, GDMO specifies how a vendor of network devices should describe products formally so that others can write programs able to recognise and manage the device. GDMO uses Abstract Syntax Notation One (ASN.1) as the rules for syntax and attribute encoding, when defining objects. ASN.1 is defined in Recommendation X.208.

■ A logical, layered architecture which defines the various layers of management, from the detailed and physical to the abstract: element, network, service, and business management. This possibly represents the first realisation that, with increasingly complex systems to manage, layers of abstraction are needed if each type of operator is to work at a different level.

Simple Network Management Protocol (SNMP)

The dominant management system used within the data networking, storage, and computing industry today, particularly in enterprises rather than backbone carriers, is the Internet Standard Management Protocol, more commonly referred to as the Simple Network Management Protocol (SNMP).

Within SNMP each attribute of a device is identified by a unique address, and a set of basic commands (get, set, get-bulk) is available to change its value. The structure of the attributes is defined in a Management Information Base (MIB), which conforms to RFC1155[†]: "Structure and Identification of Management Information for TCP/IP-based Internets." The current Internet MIB is given in RFC1213 and is sometimes called MIB-II.

SNMP version 1 was standardised in the early 1990s. This was followed by version 2 and in the late 1990s by version 3. It was originally seen as an interim and temporary stop-gap until the long-awaited CMIS/CMIP Standards emerged from the International Standards Organisation (ISO). SNMP gained wide popularity, largely because of its simplicity, and, like many things of a temporary nature, is still with us.[‡]

[†] Request for Comment: The name given to drafts and standards produced by the IETF.

[‡] "Under this proposal, on April 5, 1860, the income tax will expire" (William Gladstone, Prime Minister of Great Britain, 1853).

Recently, however, this very simplicity has hindered the development of device management—SNMP MIBs are inadequate for defining complex data and, in particular, are inappropriate for expressing relationships between attributes. Many of these shortcomings, including the lack of support for transactions (i.e. the ability of an operator to group commands so that all or none get carried out), are listed in RFC3512.

Given the extent to which SNMP management is deployed, it is essential that any new system interwork with it. For information about the interworking of WBEM/CIM and SNMP, see Chapter 14.

WBEM/CIM

WBEM/CIM emerged in the mid to late 1990s as a means of managing desktop computers. In the late 1990s it started to evolve into a more general-purpose management tool and the number of implementations started to grow.

Initially the CIM standard was defined by the DMTF and the WBEM initiative was driven through a consortium comprising BMC Software, Cisco Systems, Compaq, Intel, and Microsoft. In 1998, the WBEM work passed to the DMTF, which provides an open forum for its ongoing development and organises user and developer conferences. With its alliance partners and the WBEMSource (open source) Initiative, it also organises so-called "fusion" events, bringing together various WBEM implementations, to identify and resolve ambiguities in the standards that prevent complete interoperability.

The DMTF also intends to develop a WBEM certification or compliance programme for WBEM products. Version 1.1 of a CIM Compliance Specification was published in June 2002 and, although this currently has little content, it highlights the direction in which the DMTF wishes to travel.

SNMP, TMN, and CIM

How do the three existing management philosophies compare?

Coverage

SNMP, TMN, and WBEM do not address precisely the same ground and so it is difficult to compare them directly. Some of the differences are quite subtle, although important.

As you can see from Figure 3.3, SNMP is typically one of the management interfaces to a device, often working alongside a CLI. The SNMP representation of the managed device (the model encoded as

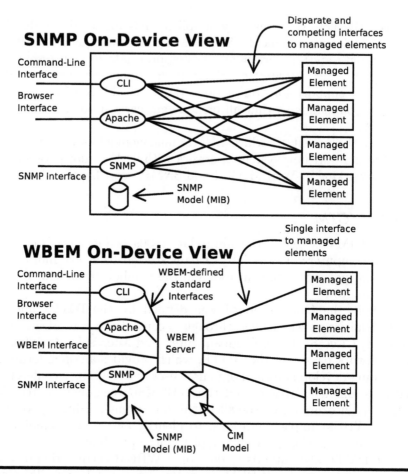

Figure 3.3   On-Device Comparison of SNMP and WBEM

a MIB) is available to the SNMP agent and represents the common language between the agent and the management workstation. It is often not, however, the underlying model used by the core management software. For very simple devices, that model may be implicit in the code; for more sophisticated devices it may be the SNMP MIB, but will more likely be a proprietary internal representation.

Although it can be used in the same way as SNMP, WBEM is most powerful when its model and server lie at the heart of the on-device software. A WBEM interface to management workstations or higher-level management devices (so-called WBEM clients) is provided out of the device. Of course, other interfaces can be supported, such as a command-line or SNMP, but these communicate with the WBEM server using the same language.

Informally, SNMP can be thought of as an interface technology that allows operators to reach a managed device. Once there, the operator's commands are translated into an internal form (often the CLI format). WBEM, on the other hand, is more than an interface; it provides the local model and software while offering a standard interface outwards and the CLI becomes just another means of access.

This has further beneficial implications for WBEM: complex systems, particularly in the telecommunications field, are managed hierarchically. A particular device is managed through an element management system, a number of devices working together in a network will be managed through a network management system and a number of networks will be managed through an operational support system (OSS). Because of the different needs, the protocols at different levels of this hierarchy have traditionally differed and an "integration tax" has been paid at each level. WBEM and its CIM models offer a possible unifying structure whereby the same models and protocols can be used all the way from the OSS to on-device management.

Orientation

The orientation of the philosophies is a major point of disagreement: SNMP is datacentric ("set the value of X," "get the value of Y"), as distinct from the task-centric orientation of WBEM ("do this," "do that"). Another pair of terms used to describe the same concepts are that SNMP is "structural," whereas WBEM is "behavioural."

TMN's Additional Network

One major point of dissent is clearly the introduction by TMN of one network (see Figure 3.2 on page 21) to manage another, raising obvious questions of recursion as we introduce a management network to control the management network. The idea probably arose from the voice-switching world where the voice network is no longer used to transfer information between switching points, this being carried over a separate Signalling System 7 network.

Although the SNMP standards do not specify whether the same or a separate network should be used for management, typically the same components are used for the network being managed and for the network over which management information is transferred.

Similarly, WBEM has nothing explicit to say about the separation or integration of these two networks, although an integrated model is tacitly assumed.

Of course, the SNMP and WBEM integrated models do not prevent a service provider from using different access mechanisms to reach the managed device: dial-up modems, in-band tunnels, etc. SNMP and WBEM do not prescribe an overlay management network.

Modelling Constructs

SNMP has a very simple modelling language, while both TMN and CIM use sophisticated object-oriented languages that permit more intuitive modelling and the ability to build larger models without them becoming unwieldy. I give an example contrasting the SNMP and CIM descriptions of an IP routing table starting on page 56.

As a simple example, consider adding a new type of interface port to a device which you manufacture. Intuitively, this new type of port belongs with other ports (Ethernet, Wireless, Token Ring, etc.). In CIM you would describe it as a child of the CIM_NetworkPort class where it is a sibling of EthernetPort, WirelessPort, and TokenRingPort classes. Not only is this natural, but it also reduces your definition workload, because your new class will automatically inherit a great many attributes (properties) from its parent: speed, port number, link technology, permanent address, etc. In addition to having all this work already completed, because of its position in the model, the new port can be associated with a whole collection of port-specific statistics, a physical connector, and a physical module on which it is implemented without the need for additional work. Using the CIM modelling technique, a model can be built quickly because much of the groundwork has already been done and the resulting model is likely to be more consistent—any external management system can make certain assumptions about all CIM_NetworkPorts, including your new one.

Manipulating the Model

Another benefit of the object-oriented nature of CIM is that it allows a suitably authorised operator to manipulate, not only the state of a component ("set port 3 to UP"), but also the state of what may be managed ("add a new association between ports of this type and protocols of that type"). This provides a very powerful tool for manufacturers and their distributors to add value to the management of a device, in essence tailoring the management to particular vertical markets. Generally, a management workstation connecting to a device over SNMP needs to know *a priori* what attributes may be manipulated. With CIM, a management workstation is able to extract information about what may be managed from the device itself.

Modelling Languages

The languages used to define models of devices also differ; SNMP uses a simple subset of ASN.1, TMN uses a superset of ASN.1, and CIM uses a much more programming-like language, *mof.*

Workstation to Model Interface

The interaction between the management workstation and the managed system also differs between the architectures. SNMP offers a few basic commands (get, get next, set, get bulk), assuming that the devices will be simple and the workstations complex. TMN offers a wider selection of basic commands (including filtering), assuming more sophistication on the managed device. CIM continues this trend, allowing functions (called methods) to be associated with particular elements. This approach is more intuitive for programmers—SNMP relies a great deal on the side effects of setting different parameters, something which can cause problems when several fields need to be updated atomically. CIM defines a method to perform the complex operation and simply invokes it. As a very simple example, consider taking an Ethernet port out of service. With the SNMP approach, the operator may set the value of a `PortRequiredStatus` property to DOWN and rely on a side effect of this setting to cause the port to go down. With the CIM approach, the operator would invoke the `goOutOfService()` method.

Although SNMP is designed to allow the configuration of a device through the side effects of setting parameter values, many equipment manufacturers have not exploited this in their management systems, relying instead on a non-SNMP CLI for configuration and using SNMP for presenting alarms and allowing queries.

Summary

SNMP, which was not intended to be an international standard, became popular by virtue of its simplicity. However, it has not been used extensively to configure complex devices, only to monitor and interrogate them. As management systems move towards service and business-level management, SNMP is no longer adequate. TMN could offer much of what is required, but has been tarred with the "unwieldy and inefficient" brush of so many telecommunications standards and is likely to be overtaken by WBEM.

Frequently Asked Questions

FAQ 3 *How do you pronounce WBEM?*

The general pronunciation is *web-em*, as in "I think that I'll WBEM across the head with this dead fish."

FAQ 4 *Are CIM models much harder to understand and to work with than SNMP ones?*

CIM models are harder to understand initially. But the rewards are higher—the language is more expressive, can handle abstractions better, integrates new products tightly with similar existing products and can handle much larger models without becoming cumbersome.

FAQ 5 *Where does WBEM/CIM stand at the moment in terms of industrial acceptance?*

I try to give some serious answers to this question in Appendix A on page 285. More lightheartedly, Figure 3.4 illustrates my observation of the lifecycle of many technologies. I sense that the hype reached its peak around 1999 when the industry press made almost daily announcements that the entire world, including my toaster, would be managed by WBEM/CIM by 2001. A period of sober reconsideration seems to have left us on the gradual upward trend that precedes either enthusiastic industry acceptance or abrupt industry rejection.

Because there appears to be no competing technology offering the same benefits as WBEM/CIM, I believe that industry will accept rather than reject it.

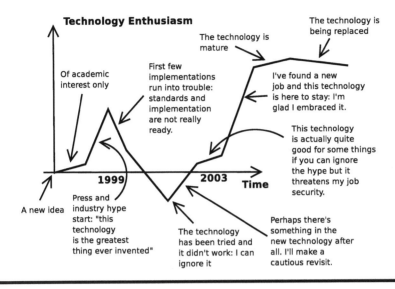

Technology Enthusiasm

The technology is being replaced

The technology is mature

Of academic interest only

First few implementations run into trouble: standards and implementation are not really ready.

I've found a new job and this technology is here to stay: I'm glad I embraced it.

This technology is actually quite good for some things if you can ignore the hype but it threatens my job security.

1999

2003

Time

A new idea

Press and industry hype start: "this technology is the greatest thing ever invented"

The technology has been tried and it didn't work: I can ignore it

Perhaps there's something in the new technology after all. I'll make a cautious revisit.

Figure 3.4   WBEM/CIM's Informal Enthusiasm Curve

STRUCTURE

STRUCTURE

Chapter 4

The WBEM Architecture

Overview

The chapter contains a description of the WBEM architecture. The WBEM server lies at the heart of this architecture and you may find it useful to place a bookmark in page 39 so that you can easily refer to Figure 4.4 as you read this chapter. I will start, however, with the problem that the WBEM architecture is designed to solve.

Figure 4.1 illustrates the management gap which WBEM is designed to fill: on the left are operators' workstations and on the right a device containing hardware (cards, slots, power supplies, etc.), software (word processors, Web browsers, OSPF routing stacks, operating systems, etc.) and services (voice mail, virtual networks, firewall, etc.). A mechanism is needed to connect the two to allow authorised operators to configure and interrogate the device and to receive information about alarms and events occurring on it. Equally important, it is necessary to prevent unauthorised operators from doing the same thing.

Note that when I use the term "operator" in this chapter (and elsewhere in this book), I mean something or someone giving commands to the management system. This may be a human sitting at a workstation, it may be a program running a script to configure many devices or it may be a higher-level management system. By virtue of its sophisticated modelling language and flexible interface, WBEM is particularly suited for use in hierarchical management systems like this.

The two sides of Figure 4.1 represent two different levels of abstraction. The state of the cooling fan within the hardware of the device might actually be accessed by reading the top two bits of register 17 in the integrated circuit at address 0xFFFF6638 and comparing the result

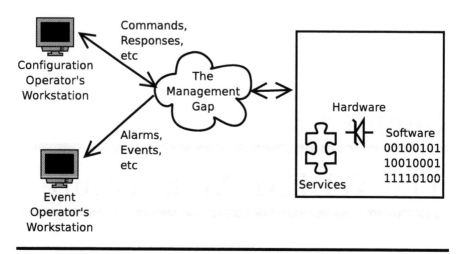

Figure 4.1   The Management Gap

with 00 (fan failed), 01 (fan operating), 10 (fan functional but off), and 11 (fan not fitted). This is something that the operator would prefer not to know; he or she simply wants to know the state of the fan, not how to obtain that state. We need a means of hiding the details from the operator, particularly if two different devices are being managed at the same time, each with a fan, but with disparate ways of accessing their state.

In Figure 4.2, a WBEM server, as illustrated in detail in Figure 4.4, has been added. The primary role of the WBEM server is to act as a broker between the WBEM clients, working on behalf of the operators, and the providers, working with internal knowledge of the hardware, software, and services. It can act as a broker in this way because it offers standard interfaces to these components—thereby isolating the clients from the providers. Clients need have no knowledge of how their requests are being handled; in fact they need have no knowledge of the existence of providers. Providers, on the other hand, need have no knowledge of the origin of the commands they are being given to execute. The WBEM server provides this isolation, and WBEM clients and providers should be designed to be independent of each other to avoid restricting enhancements to the management system. This isolation means that a WBEM client, needing to know the amount of free disk space on a computer, should be able to address precisely the same request to PowerPC-based PCs running Linux, Sparc Computers running Solaris, eServers running zOS, RISC computers running HPUX, or Intel-based PCs running Windows, and get an answer in the same format. The actual mechanism for obtaining the information on each

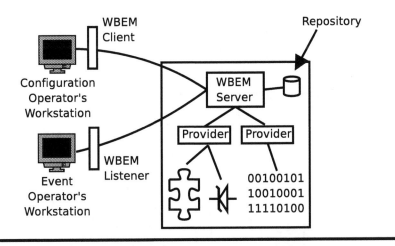

Figure 4.2   The WBEM Components

of those operating systems may vary, but the means of accessing it does not.

The protocols used between a WBEM client and a WBEM server are described in Chapter 7.

Of course, the abstract model of the device must find its way from the imagination of the designer into the WBEM server. The starting point for this process is the set of models standardised by the DMTF and published in a formal language known as *mof*. For some devices, these standard models may suffice, but if they do not, you can extend them by writing additional *mof* code, known as "extension models." The standard and device-specific *mof* code is then passed through a compiler (see Figure 4.3) which acts as a WBEM client, and loads it into the WBEM server.

In addition to checking the syntax of the *mof* code and producing a version of the model suitable for the WBEM server, the compiler may produce other output—including skeleton code for providers and user documentation. This varies from implementation to implementation.

Therefore, the steps that you need to take to manage a new device using WBEM/CIM are as follows:

1. Examine the DMTF's standard models and determine how well they meet your needs. Write any additional *mof* code to extend the standard models as required. I describe the *mof* language in Chapter 5 and the standard models in Chapter 6.

2. Obtain, compile, and integrate a WBEM server with the software on your device. There are various open source and commer-

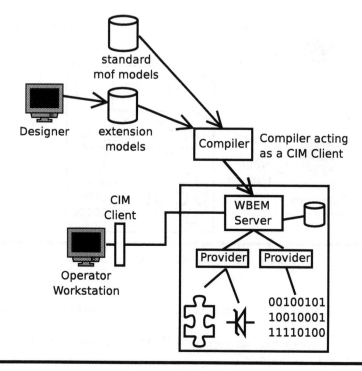

Figure 4.3 *mof* **Compilation**

cial WBEM servers available in C, C++, and Java (see Chapter 15).

3. Write the providers necessary to handle the low-level detail of your device. The WBEM architecture gives a useful infrastructure into which the providers are built, governed largely by the interface between the WBEM server and the providers (see Chapter 11 and the example in Chapter 12).

4. Write the necessary clients to interact with the WBEM server on behalf of the operators. Client development is also usefully circumscribed by the interface with the WBEM server (see Chapters 7 and 8). Note that the actual interface presented to the operator (graphical, textual, etc.) is outside the scope of WBEM/CIM; the WBEM/CIM architecture stops at the client. In effect, the use of the WBEM server as a broker allows the development of clients and providers to proceed independently of one another.

Once this work has been carried out, the management system is ready for use. For most applications, the most significant savings in development time arise through the reuse of the standard models and the WBEM server code. A substantial amount of work has gone into producing both of these and ongoing enhancement and bug fixing is also occurring. Given the free availability of the DMTF models and the open source WBEM server implementations, some decisions that you need to make concern your involvement with the communities producing them:

- Would it be useful to take your extension models back to the DMTF for standardisation within your industry? Would you get commercial advantage out of using a standardised management interface or do you need to keep it proprietary?
- Should you contribute to the development of the open source WBEM servers? By doing so, you can better influence the direction and release timings of the development.

Structure of the WBEM Server

The components of the WBEM server are illustrated in Figure 4.4. The WBEM server is a broker between various independent elements: primarily the WBEM clients and listeners which represent the operators or higher-level management systems and the providers which interact directly with the actual hardware or software being managed.

The heart of the WBEM server is the CIM Object Manager (CIMOM), which uses information stored in a model in the repository to direct commands and responses between the WBEM clients, providers, and listeners. The CIMOM provides interfaces to:

A repository, in which an abstract model of the device's managed characteristics (hardware, software, and services) are stored. I describe how to build such a model in Chapter 5.

Providers, which act as drivers and interface between the abstract world of the model and the messy characteristics of real hardware and software. The providers supply the intelligence to translate an abstract request ("return the state of the fan") into a specific command ("read the top two bits of register 17 in the integrated circuit at address 0xFFFF6638 and convert the result in accordance with the following mapping, etc."). Providers are discussed in detail in Chapter 11.

Clients, able to interact with the CIMOM to manipulate the model as requested by an operator. Clients may exist on the device being managed, on the operator's workstation, or on some other device; see Chapter 13.

Listeners, which receive information about events and alarms that have occurred in the managed device and about which an operator has expressed interest.

Security plug-ins, which typically interface to external authentication services to authenticate users and operators.

Architectural Options

In Figure 4.3 I have made the tacit but perhaps typical assumption in allocating components to devices; I have assumed that the client code *will not* be implemented on the device being managed, but probably in the operators' workstations and the WBEM server and providers *will* be implemented on the device. Although this is typical, it is not required by the standards—they are silent on the subject of the distribution of components.

Where the external operator interface is other than WBEM (for example, a command-line interface or SNMP), the WBEM client may need to reside on the managed device, as shown in Figure 3.3.

For other devices it might be impractical to have the WBEM server and repository running on the managed device because of memory or processing constraints. Then it might be necessary to have these running on a management workstation or intermediate computing device with the providers running on the managed device remotely. In this case, the WBEM server could interface with "Provider Proxies," which themselves interact with the real providers over some form of communications infrastructure. Most of the available WBEM server implementations either support the concept of remote providers or have declared the intention of doing so shortly. The emerging standard specification for the interface between the WBEM server and its providers (CMPI) explicitly allows for a remote connection to the providers. If you are using a WBEM server which does not support them, then producing a home-crafted interface using a technology such as an embedded ORB would probably not be too difficult.

These two architectures are illustrated in Figure 4.5, where Architecture A has the conventional distribution with the WBEM server on the managed device and Architecture B illustrates the condition when the managed device cannot contain the WBEM server. Architecture B could potentially lead to significant problems if, as I have heard proposed, the WBEM server was actually implemented on each operator's

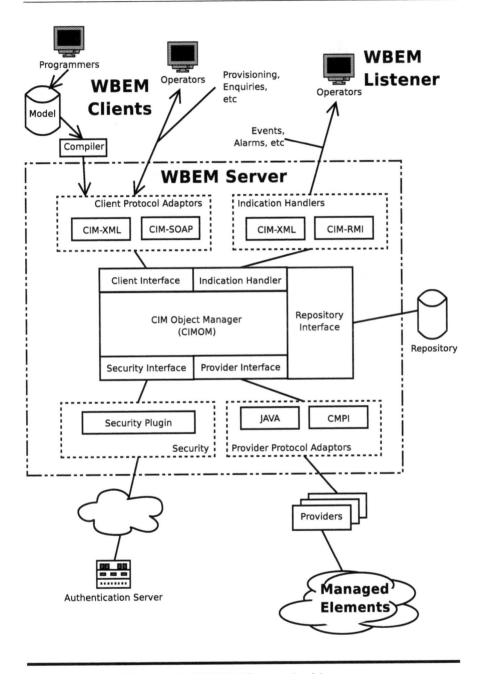

Figure 4.4   WBEM Server Architecture

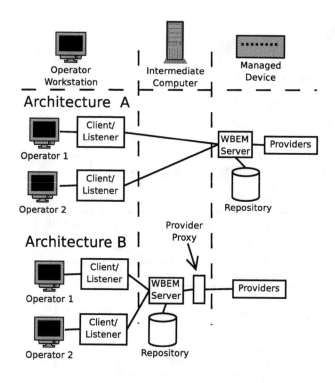

Figure 4.5   Different WBEM Server Locations

workstation. Without co-ordination between the different workstations (something that would be difficult to achieve), the alterations made to the structure of the information held in the repository would not propagate to other workstations. If co-ordination *were* built in between the workstations, then the failure of the "master" workstation could be difficult to handle. Bear in mind that a workstation is normally the least reliable component of the systems illustrated in Figure 4.5. It would also be difficult to co-ordinate the access to the remote providers between the various WBEM servers.

Even if the WBEM server *can* be executed on the device being managed, this may be an inappropriate location for a model which defines services across several devices. In this case, a hierarchy of WBEM servers may be more appropriate, with a high-level server containing the model of the interdevice services and communicating with low-level servers that hold device-specific models. This is illustrated in Figure 4.6 where the communication is performed by providers of the high-level

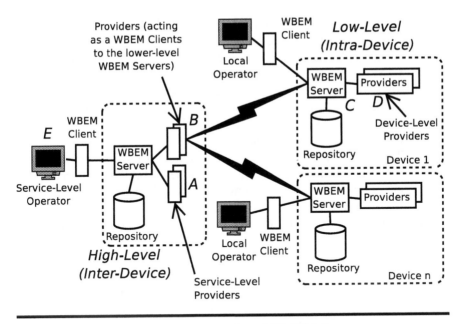

Figure 4.6   A Hierarchy of WBEM Servers

model (B) acting as WBEM clients to the lower-level WBEM servers (C).

The ability to use the same abstract model at different levels of management like this is unique to WBEM, being possible neither with SNMP nor TMN. Using a single protocol with commands referring to a common model removes the "integration tax" normally met in a hierarchical management system.

Example

I describe the interfaces between an operator, a WBEM client, a WBEM server, a listener and a provider in detail in later chapters. To give you an informal view of the interactions, this section outlines a simple use-case of a service-level command entered by an operator at position E in Figure 4.6.

Assume that the service-level operator has entered a command regarding the current state of a particular instance of a service. This operator may have little knowledge or understanding of the underlying components of the service which are implemented in a heterogeneous collection of devices from different manufacturers. The sequence of events is something like the following.

1. The operator's command is translated from the specific user interface (perhaps command-line or graphical) into a query on a particular service status. The service has been modelled in CIM and its model is contained in the repository of the high-level management server shown in Figure 4.6. The current state of the service has been modelled as a variable which may take various values such as "not installed," "running," "down for maintenance," and "failed."

2. The high-level WBEM client constructs a request for this value using a standard format defined by the DMTF. This request is encoded in a protocol known as CIM-XML and passed to the high-level WBEM server using HTTP.

3. The high-level WBEM server consults its model stored in the repository and finds that a provider is associated with the requested value.

4. The high-level WBEM server therefore invokes the service provider (marked *A* in Figure 4.6) with a request for the current state of the service. Unfortunately, this is not something which the provider monitors in real-time because this would require a significant amount of unnecessary traffic—if no one wants to know about the value, why collect it? The provider does, however, understand the semantics of the service and knows how to collect the information.

5. The service provider has access to the high-level model and, by acting as a WBEM client, extracts the information from the WBEM server about the precise devices forming part of the service. (Note: It could have held this information locally; but because the WBEM server is designed for this type of function, this is often unnecessary).

6. The service provider then requests the values of the current state fields from each of the component devices from the high-level WBEM server. The high-level WBEM server consults its model and finds that there are providers associated with each of these values and invokes them—the providers marked *B* in Figure 4.6.

7. These providers do not have the information stored locally, but they understand how it may be retrieved from lower-level management systems—those on the devices. They therefore act as WBEM clients and issue commands, encoded in CIM-XML, to the individual low-level WBEM servers (marked *C* in Figure 4.6).

8. Each low-level WBEM server follows a pattern similar to that of its high-level brother: it scans its model (now very device

specific) and finds that the information it requires is handled by a provider. It therefore invokes a provider (D in Figure 4.6).

9. The low-level providers consult the actual device as necessary (reading registers, etc.) and return a response to the WBEM server.

10. The low-level WBEM servers each respond to the calls made to them by the higher-level system and eventually all of the responses are gathered together by the high-level service provider A.

11. The service provider then does whatever calculation is appropriate for the overall state of the service, given the states of the component parts, and returns the value to the high-level WBEM server.

12. The high-level WBEM server encodes the response in CIM-XML and passes it back to the WBEM client, which converts it into the required format for the operator and displays it.

When following this example, note that each provider only handles what it understands directly, the isolation provided by the WBEM architecture having allowed the management problem to be partitioned into smaller, more tractable parts. Note also that all messages sent between the WBEM clients and WBEM servers are encoded in exactly the same format, independent of the hierarchical level—the commands and responses are always couched in terms of the model and the model is available to all parties.

The outcome of this rather complex example is that the operator entered a command at the service level, the level he or she understands, and got back a response at the same level. The management systems on the individual devices received only commands at their level and they responded at that level. There is a clear demarcation between the levels, but a common model, language, protocol, and architecture provided by WBEM to allow seamless interaction.

Frequently Asked Questions

FAQ 6 *Figure 4.2 on page 35 shows a client and a listener. What is the difference?*

The WBEM architecture differentiates between two types of management:

1. Operators entering commands to configure items and enquire about their status.
2. Operators receiving information about events and alarms which have occurred in the system.

The major distinction between these two is in what actually starts the interaction between the operator and the managed system. In the first case, the operator starts the interaction; in the latter, an event occurring in the system is the original stimulus.

The software which acts on behalf of an operator entering commands is known as a WBEM client. The software acting on behalf of an operator waiting to receive details of events and alarms is called a listener.

FAQ 7 *I understand that the WBEM server acts as a broker between the clients and providers, but how much checking does it do? Does it, for example, check the types and ranges of properties in an instance that a client is creating before it passes those parameters to the provider?*

Although this question is reasonable, given the content of this chapter, to understand the answer you unfortunately need more information about the way a CIM model is constructed. I have therefore replicated this question as FAQ 20 on page 84.

FAQ 8 *If I use an open source WBEM server in a product that I sell, will legal issues arise?*

Open source code is normally copyrighted and released subject to a licence. It is essential that you (and your lawyers) read those licences very carefully. Generally, the open source licences range from "here's the code, do what you want with it, but don't prosecute us if it doesn't work" to the GNU General Public Licence (GPL), which is long and complex. Using open source software in a product that you are selling is easy and legal to do, but you (and all your programmers) must be aware of the licensing issues and treat them with respect. openPegasus is one of the better open source C++ WBEM servers and its licence is reproduced on page 313 of this book. As you can see, it is effectively of the "do what you want but don't prosecute us" type.

FAQ 9 *What type of database is used for the repository? Do I need to use a commercial database for the repository?*

The WBEM/CIM specifications mandate no particular type of data storage for the repository; each WBEM server implementation is different. Although each WBEM server implementation comes with a default repository, the implementations generally try to provide a clean interface so that any type of data store can be used; perhaps flat files on disk, perhaps a sophisticated database, perhaps linked lists in memory.

openPegasus, for example, comes with a very simple (and very inefficient) repository comprising a disk file for each class, the content of each being stored in XML. This is very useful for debugging as the repository files can be browsed using a normal text editor, but it may be too inefficient in space or processing time for a production system.

Chapter 5

CIM and *mof*

The Concept of a Model

Creating a language to express their ideas has been a challenge to scientists and engineers since the beginning of time—the Romans were handicapped in their use of arithmetic by having a nonpositional number system unable to express zero. Because their representation was poor, they were held back in expressing their ideas.

We need to select a language in which to express the concepts for managing devices and services. The language must be rich enough to include the complex and abstract relationships between items, but be simple enough for software to handle efficiently. The idea of a position-based number system with a zero evolved gradually and, similarly, our languages for expressing management models are also changing and refining.

Rather than expressing arithmetic, our linguistic aim is to define a model of the devices and services we intend to manage. In this book I have studiously used the term "model" to refer to the description of the things being managed. You may be familiar with the terms "data model" and "information model." I have avoided these because the distinction between them is, at best, the source of quasireligious debate: in January 2003 the Internet Engineering Task Force (IETF) even issued a Request For Comment (RFC)—number 3444—entitled "On the Difference between Information Models and Data Models." The preface to that document reads:

> There have been many discussions to understand the advantages and disadvantages, as well as the main dif-

ferences, between various languages. For instance, the
IETF organized a BoF* on "Network Information Modeling" (NIM) at its 48th meeting in 2000. During these
discussions, it turned out that people had a different
understanding of the main terms, which caused confusion and long arguments. In particular, the meaning of
the terms "Information Model" (IM) and "Data Model"
(DM) turned out to be controversial.

In an attempt to address this issue, the IETF Network
Management Research Group (NMRG) dedicated its 8th
workshop (Austin, December 2000) to harmonizing the
terminology used in information and data modeling. At-
tendees included experts from the IETF, DMTF, and
ITU, as well as academics who do research in this field.
The main outcome of this successful workshop – a better understanding of the terms "Information Model" and
"Data Model" – is presented in this document.

Having read the document several times, I confess to feeling that I
still do not understand the difference but that, luckily, it probably does
not matter.

So, what do I mean by a model? Figure 5.1 illustrates the concept: somewhere is stored the actual information which makes a device
operate correctly. This is shown on the right-hand side of Figure 5.1
and it may be stored in registers in an integrated circuit, in computer
memory, etc. It may be stored in one place or be distributed.

On the left-hand side of Figure 5.1 I have shown a number of different operators. They have different mental models of the device that
they are managing, and it may be appropriate for the management
system to reflect their particular mental model even if it is wrong or
incomplete. I have spent my life working with software; my wife has
not. When the software running on her computer displays an event on
the screen (e.g., when Netscape has failed yet again), she immediately
and unwillingly becomes a management operator. It is important that
the message she receives fits with her mental model of what is happening inside the computer. Unfortunately, most software having been
written by people like me, it rarely does. This increases the chances
that she will react inappropriately to the message—perhaps powering

---

* Birds of a Feather Meeting: i.e., a meeting of people interested in the topic, but not
forming part of the standardisation process.

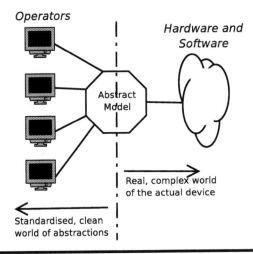

Operators

Hardware and Software

Abstract Model

Real, complex world
of the actual device

Standardised, clean
world of abstractions

Figure 5.1   The Role of the Model

the computer off and on again, compounding the original problem by corrupting the file system.

If the operator is in the business of scheduling the repainting of telecommunications street cabinets, then his or her mental model of the device is of a metal cabinet of a certain colour, last repainted on a certain date. The interior of the cabinet is probably irrelevant. If, on the other hand, the operator's responsibility is to configure and operate the IP router within the cabinet, then he or she probably never pictures the cabinet, let alone its colour.

The model stands between these different operators' mental models and the normally messy, real representation of the information. It tries to be sufficiently abstract and consistent to support different viewpoints, while being sufficiently close to the real devices to be efficient. In general:

■ The model should be expressed in a formal language so that it can be manipulated by a computer. Preferably it should also have a graphical representation to allow humans to understand it more easily.

■ The model must be able to meet the needs of all potential management operators—in the previous example it must encompass both the colour of the cabinet and the IP addresses of the router. In this case, the management system will probably need to confine a particular operator to one part of the model; it might not be appropriate for the painter to be able to read or change the

IP addresses or for a bored IP operator to schedule the cabinet to be repainted. This involves providing different windows onto the same information: the operators can then choose the window through which they look.

■ The model must be able to express high-level (service) abstractions. An operator configuring an end-to-end service (such as a Virtual Private Network†) need not be exposed to the internal details of every port on every device—the model should support the concept of a Virtual Private Network and allow the operator to work with this abstraction, the translation to ports on devices being done within the management system.

■ The model should allow an operator's workstation to enquire about and manipulate the structure of the model itself. Management workstations can be simpler if they can ask the model not only for details of particular entities ("is port 27 on card 11 in service?"), but also for details of the model itself ("with what types of thing is port 27 associated?"). This allows the management workstation to find out what it can ask the device—it does not need to know this *a priori*—enabling the development of more generic management workstation software.

Another advantage of this self-describing model is realised if its structure can be changed by an approved operator while the system is running: an operator can change not only the status of a port from *disabled* to *enabled*, but also the linkages in the model relating to the port (which modules need to be informed when it fails, etc.). This could, for example, allow a company which buys and distributes devices to tailor management features for a particular market without involving the original manufacturer. The concept of the "approved operator" is essential here: a malicious operator could do great damage to a system if given this level of control: the system needs to address questions of operator authentication (are the operators who they say they are?) and authorisation (what may these operators do?).

■ The model should support the modelling of collections of entities and associations between entities where these are relevant to the application. An end-to-end link in a network, for example, might be modelled as a collection of point-to-point links. For some applications it might be more consistent with the operator's mental model to work with the end-to-end link.

---

† A self-contained network dedicated to one customer and using a subset of the resources of a real network.

Modelling Terms

Most professions have taken words from the public domain and given them different meanings. When a mathematician and a wedding organiser use the phrase "a function in a field," for example, they probably mean very different things. Miscommunication increases when a "ring" is also thrown into the conversation.

The modelling world has also taken its jargon from common words and attached meanings related to, but not precisely the same as, the common usage. The next few paragraphs discuss the use of the terms class, inheritance, instance, property, method, association, indication, and object.

Class

The concept of class is fundamental to everything else. A class is an amalgam of characteristics defining a particular group of things (note my careful avoidance of the loaded word "object"). *Domestic-Dog*, for example, is a class. It does not mean any one particular dog (Fido or Patch), but describes a whole collection of dogs. There are characteristics that all domestic dogs could be assumed to have: a name and an owner, for example.

Class Inheritance

A class can have subclasses (also themselves classes). The class *Dachshund*, for example, might be a subclass of *DomesticDog*. A dachshund will have all the characteristics of a domestic dog (name, owner, etc.), but also some useful or important characteristics unique to itself: its clearance, for example, between belly and ground, measured in millimetres.

We say that the class *Dachshund* is a subclass of the class *Domestic-Dog* and that it inherits from *DomesticDog*. This concept of inheritance is important as it means that, when we define the class *Dachshund*, we do not need to repeat that dachshunds have a name and an owner; those characteristics are inherited from *DomesticDog*.

Subclassing in this manner is also sometimes called specialisation: *Dachshund* is a special type of *DomesticDog*.

Instances

Of course there are also actual dogs which belong to the *Dachshund* class: Fido, for example. Fido is said to be an instance of the class

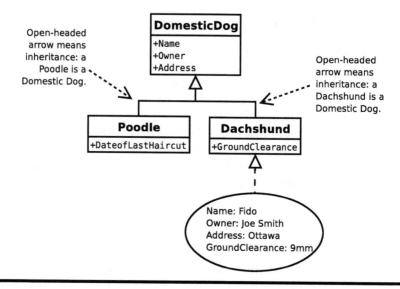

Figure 5.2   Fido without an Association

Dachshund. Because *Dachshund* is a subclass of *DomesticDog*, Fido is also, by inheritance, an instance of the class *DomesticDog*. Fido, therefore, can be expected to have a name, an owner and a value for his ground clearance.

It is important to notice the chain of "is-a" phrases: Fido *is a* Dachshund which *is a* DomesticDog. I explore the crucial difference between *is a* and *has a* in more detail on page 56 and in Appendix B because this is one of the major characteristics of CIM modelling and it represents a significant departure from the way in which SNMP models are produced. Figure 5.2 illustrates these relationships.

Properties

We have spoken of the characteristics of a class (e.g., name, owner, ground clearance). In CIM-speak, these are called properties. So, correctly, an instance of a class has the properties defined in its class and all of the classes from which its class inherits.

Methods

A method is another characteristic of a class: it represents a function that an instance of the class can be asked to perform. An instance of the *DomesticDog* class, for example, might know how to carry out the commands "sit!" and "beg!"

More usefully, a *CIM_PrivilegeManagementService* which, as its name suggests, holds information about the privileges of a particular user, can be asked to "AssignAccess" or "RemoveAccess" rights for a particular user. These functions are methods of the class.

Similarly, the *CIM_LogicalDevice* class, which represents the management information for a device, can accept the command "Reset!"—this is therefore a method of the *CIM_LogicalDevice* class.

Associations

Continuing our dog example, we need to specify Fido's owner. We could do this by making the owner's name and address properties of a *DomesticDog*, and then set these values in Fido's instance. This is illustrated in Figure 5.2 where it can be seen that Joe owns Fido.

This technique would, however, be inefficient if Joe had several dogs: Joe's name and address would have to be repeated and, if his address changed, every occurrence of it would have to be found and modified. It would be better to store the owner Joe as an instance of a class *DogOwner* and then somehow link Fido's instance to it. This is illustrated in Figure 5.3. In CIM-speak, we say that we have "created an association" between the *DomesticDog* class and the *DogOwner* class.

Figure 5.3 contains an instance of a class called *DogOwnership*, which *is an*‡ association. The instance of the association links Joe's instance with Fido's. This separation of the association from either Fido or Joe, making it a separate entity, is another important characteristic of CIM, giving it its flexibility and extensibility. This association effectively adds *has a* into our model: an instance of *DogOwnership* says that a particular *DogOwner has a* particular *DomesticDog* or, equivalently, that a particular *DomesticDog has a DogOwner*.

The "pointers" held in the association, pointing to the associated items, are known in CIM-speak as references.

An important characteristic of associations as created as in Figure 5.3 is that the *Dachshund* and *DogOwner* classes have no cognisance of any association between them: associations can therefore be added to a running system without disturbing the classes already created.

Another important characteristic of associations is also illustrated in Figure 5.3; although the *DogOwnership* association is defined as being between a *DomesticDog* and a *DogOwner*, it is automatically inherited by subclasses of these. Thus an instance of a *Dachshund*, because it is also by inheritance a *DomesticDog*, can also have a *DogOwnership* association.

---

‡ OK, it is not always *is a*, sometimes it becomes *is an*.

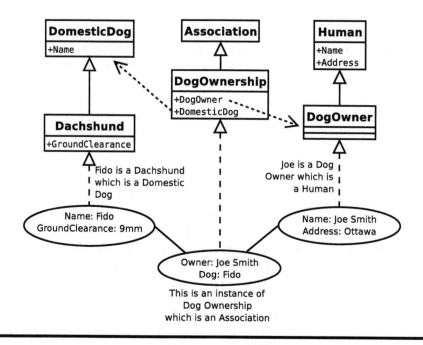

Figure 5.3 Fido with an Association

Note that, in addition to "normal" associations, CIM also contains the concept of a weak association. Because this type of association requires you to know about the concept of keys, I have deferred discussion of it until page 68.

Indications

We have nearly completed our vocabulary lesson—there is just one more major topic: indication. The terms covered so far (class, instance, association) allow us to describe the dog and its properties (ground clearance and owner). There is a further concept needed to describe the signals which the dog may wish to pass to us. It may, for example, bark to indicate that it is hungry. Signals arising spontaneously from instances of a class are known as indications: the dog may indicate that it is hungry by barking. Similarly, a telecommunications link may signal that it has failed by raising an alarm.

In the same way that an association is just a special type of class, so is an indication. The indication class describes a type of indication that a managed object may raise ("I'm hungry," "I have failed," "I'm on fire") and an instance of the indication is created when that event occurs.

Outside the world of dogs, events might be the failure or recovery of a component (in which case the indication would generally be called an alarm), the creation or deletion of a class or instance by an operator or a set of accounting records being passed spontaneously from the device to its management system.

I treat the concept of an indication in Chapter 8: suffice it to say that when an alarm occurs on a device with CIM management, where any operator or logging system has registered an interest in knowing about that particular type of alarm, an instance of an indication class is created describing the alarm, giving details of its severity, date, and time of occurrence, etc. Programs known as handlers may be registered with the WBEM server to receive copies of Indications meeting certain criteria ("severity at least CRITICAL and temperature greater than $32°C$"). When an indication is signalled, the WBEM server checks to see which, if any, handlers have registered an interest in receiving a copy and passes the information to them.

Objects

After the careful descriptions of the terms class, association and instances given previously, the term *object* is a bit of an embarrassment. The term is commonly used as a synonym for "instance," but actually may refer either to a class or to an instance. The distinction is normally not important, but the intrinsic operations on associations (see page 136), unlike other intrinsic operations, can be applied to either classes or instances. It is useful to be able to say that they can be applied to "objects."

A CIM Example

Of course, CIM is not used primarily for modelling domestic animals. Taking an example from the standard network model produced by the DMTF, consider a class for describing Ethernet ports eponymously called CIM_EthernetPort, as illustrated in Figure 5.4.

It is easy to imagine an instance of this class: a real Ethernet port on a network device. The class CIM_EthernetPort inherits from CIM_NetworkPort, which in turn inherits from CIM_LogicalPort. Thus a particular instance of an Ethernet port has all of the properties defined for the CIM_EthernetPort class (e.g., maximum data size) but it also inherits all of the properties of the CIM_NetworkPort class (e.g., whether or not auto-sensing is possible) and all the properties of the CIM_LogicalPort class (e.g., maximum speed).

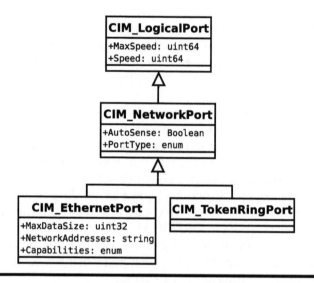

Figure 5.4 The class CIM_EthernetPort

Notice again the concept of *is a*. Ethernet port 23 on my computer *is a* CIM_EthernetPort, which *is a* CIM_NetworkPort, which *is a* CIM_LogicalPort. I have also shown CIM_TokenRingPort in Figure 5.4—it *is a* CIM_NetworkPort as well.

"Is-A" and "Has-A" Relationships

You may be familiar with modelling in SNMP. Modelling in CIM is fundamentally different because its basic means of linking one component with another is through *is a* (Joe *is an* instance of a DogOwner, which *is a* Human) rather than through *has a* (Joe *has a* dog); the *has a* relationship being modelled, where needed, by means of an association.

The difference may superficially appear small, but it actually turns the model on its head. Appendix B gives more examples and a more complete explanation, but in this section, I try to explain its impact on the model. You can think of the *has-a* abstraction as being structural ("of what is this entity composed?") and inheritance as being subtyping. By contrast the *is-a* abstraction is behavioural ("what does this entity do, what behaviour does it have?") and its inheritance is subclassing.

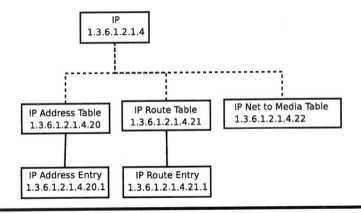

Figure 5.5   Part of the SNMP MIB-II Subtree

To make this more concrete, start with the example of an Internet Protocol (IP) table[§] and the way it is described using a *has-a* system such as SNMP and an *is-a* system such as CIM.

Figure 5.5 illustrates some of the subdivisions of IP within the standard SNMP MIB-II. IP itself is addressed as 1.3.6.1.2.1.4, meaning that it is the fourth item in 1.3.6.1.2.1, within the SNMP MIB-II. This is itself the first sub-item of 1.3.6.1.2, which is known as IETF Management. This process continues to the top of the tree, which is known as 1 or "ISO Assigned Object Identities." The part of the SNMP tree below IP, as shown in Figure 5.5, contains the IP address table (which holds the IP address of every port on the device) and the IP routing table (which holds the information to route packets, a definition of the next hop for each possible destination). This is a *has-a* relationship: IP *has an* IP address table which *has-an* IP address entry. (Note: I have dotted the upper links in Figure 5.5 because the containment hierarchy is not always strictly applied above tables.)

Contrast this with Figure 5.6, which contains that part of the CIM model concerned with IP routing tables.

A CIM_NextHopIPRoute *is a* CIM_NextHopRoute, which *is a* CIM_ManagedElement. Similarly a CIM_IPProtocolEndpoint *is a* CIM_ProtocolEndpoint, which eventually *is a* CIM_ManagedElement. Notice that a CIM_IPProtocolEndpoint is a peer of a CIM_TCPProtocolEndpoint and a CIM_LANEndpoint and (not shown in Figure 5.6) a CIM_BGPProtocolEndpoint: this is not unreasonable—they are all protocol endpoints and therefore all share some properties such as a

---

[§] The precise meaning of this is not important at the moment—assume that it is a collection of tables used in an IP router to decide how to handle received packets.

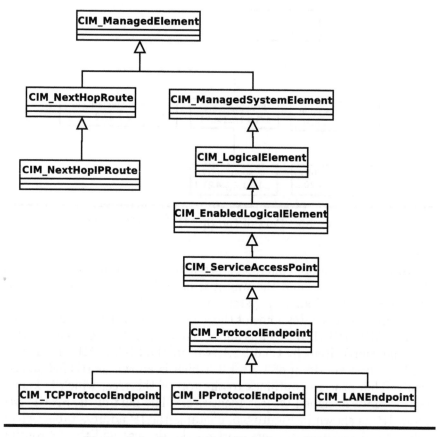

Figure 5.6   Part of the CIM Network Model

protocol type (IPv4, IPv6, IPX, AppleTalk, DECnet, SNA, etc.) and some associations (e.g., with a service), which they can inherit from CIM_ProtocolEndpoint.

This intrinsic similarity between IP and TCP as service endpoints is lost in the SNMP representation because TCP appears in a totally different subtree: rooted at 1.3.6.1.2.1.6. In the CIM model, any behaviour associated with all protocol endpoints can be encapsulated in one place: in CIM_ProtocolEndpoint. Any changes affecting all endpoints can then be made in one place.

UML for CIM

I use examples from the DMTF's standard models in this section to show their UML representation. Although I

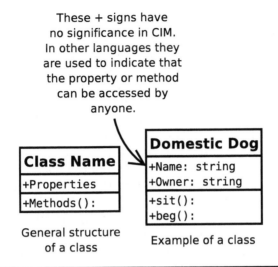

These + signs have
no significance in CIM.
In other languages they
are used to indicate that
the property or method
can be accessed by
anyone.

Class Name
+Properties
+Methods():

General structure
of a class

Domestic Dog
+Name: string +Owner: string
+sit(): +beg():

Example of a class

Figure 5.7   UML Representation of a Class

*give some explanation, it is not necessary for you to un-
derstand fully what the models mean—just to understand
the UML nomenclature that they are illustrating.*

A model can be expressed in CIM in two ways: graphically using
the Unified Modelling Language (UML) or textually using a language
called *mof*. I describe each of these representations in the next sections.

The Unified Modelling Language

The Unified Modelling Language (UML) is a graphical language standardised by the Object Management Group (OMG) which is used to
"visualise, specify, construct, and document" the structure of a system. There are many books about UML and the Object Management
Group's Web site (http://www.omg.org) contains links to a number of
tutorials. In fact, the amount of UML that you need to read and write
CIM models is so small that if your only reason for learning UML is to
read and write CIM, then this chapter should suffice.

UML for Classes and Subclasses

We have already seen examples of classes in UML notation—domestic
dogs and IP routing tables. Figure 5.7 shows the general structure of
a class in UML: a rectangle divided into three parts:

1. The highest subdivision contains the name of the class. The name may be written in italics to indicate that that class cannot have instances: it can only be a superclass of other classes.

   Within CIM, a class name has structure, being composed from a so-called schema name, an underscore, and a class name. A schema is another word for a (sub-)model and so, for example, all of the CIM classes are of the form: *CIM_classname,* where CIM is the schema and classname, reasonably, the class name. If you create a sub-model for your company's products, to avoid potential clashes with the names of other classes which you may inherit, you should use a trademarked name as the schema name.

   If, for example, you work for the ACNE Manufacturing Company, then you might choose the term ACNE for your schema. In that case any classes you create would have a name with the format ACNE_classname. Of course, such a class may inherit from a class in the CIM schema.

2. The middle subdivision lists some or all of the properties that characterise the class.

3. The lowest subdivision lists some or all of the actions ("methods") that the class supports (i.e., the functions it can perform). The right-hand side of Figure 5.7 shows the example of the *DomesticDog* class taken from Figure 5.2 on page 52.

The next notation is that of linking classes together to indicate that a subclass *is a* particular class. This is illustrated in Figure 5.2: simply an open-headed arrow joining the subclass to the class. That figure shows that a Dachshund *is a* DomesticDog (as *is a* Poodle).

UML for Associations

Although UML itself is a very rich language with symbology for many types of diagrams, there are only a few more notations you need to understand when reading and writing CIM diagrams: those for the various types of association.

Although an association is a class in CIM, it is generally not drawn using the rectangular box that designates a class; it is rather shown as a simple line between the classes which it associates. Figure 5.8, for example, taken from the DMTF's User Model, shows that an instance of the class CIM_SecurityService (which *is a* CIM_Service, which *is a* CIM_LogicalDevice) may be associated with an instance of the class CIM_System through an instance of the association CIM_SecurityServiceForSystem. What this means is that a particular CIM_System may be associated with a particular CIM_SecurityService.

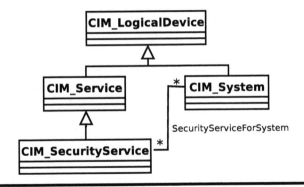

Figure 5.8   UML Representation of an Association

Another point to note in Figure 5.8 is the star appearing at the end of the association line. This indicates that any number from 0 upwards may be part of the association. For example, one CIM_System may have any number of CIM_SecurityServices, including none. Also, one CIM_SecurityService may be associated with any number of CIM_Systems. Other symbols which could appear at the end of an association include "1" (i.e., exactly 1) and "3..5" (between 3 and 5 inclusive).

Because an association is, in fact, a class within CIM, it is occasionally necessary to indicate that it has properties other than the references to the things it associates. These are drawn under the association's name or close to the association's line and connected to it by a further line. An example is given in Figure 6.8 on page 109 where CIM_PolicyConditionInPolicyRule has two attributes: GroupNumber and ConditionNegated.

UML for Aggregation

Another notation which is sometimes useful is "aggregation": this is the extension of the *has a* concept in the *is a* model because aggregation is just a special type of association. Figure 5.9 illustrates a particular aggregation taken from the standard CIM device model.

The lines with the open diamonds indicate aggregation, with the diamond being at the "contains" end. Figure 5.9 shows that a CIM_DiskGroup may contain (*has a*) an aggregation of CIM_StorageExtents.¶ It may also contain an aggregation of CIM_Disk-

---

¶ The term StorageExtent describes a collection of storage devices (disks, etc.) used together to create a single logical storage device.

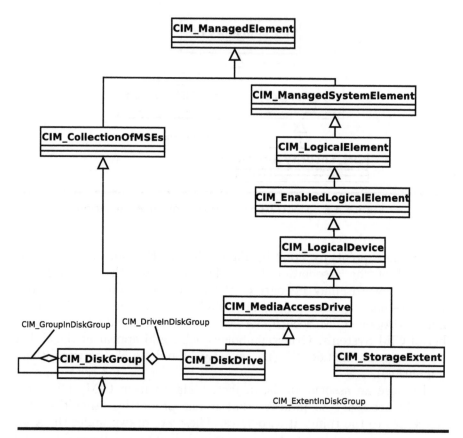

Figure 5.9   UML Representation of an Aggregation

Drives. It may also, recursively, contain an aggregation of itself, of CIM_DiskGroups.

Figure 6.8 on page 109 also uses an "aggregation diamond": in this case, to show that a CIM_PolicyRule (never mind what this is at the moment, have faith that all will become clear later) aggregates CIM_PolicyConditions and CIM_PolicyActions.

You will see later in this chapter that classes, methods, and properties may have so-called qualifiers attached to them. These qualifiers typically give additional information about the class, method, or property, restricting, for example, a numerical value to be less than 42. There is a special qualifier, **AGGREGATION**, which may be added to the *mof* definition of an association to indicate that it represents an aggregation.

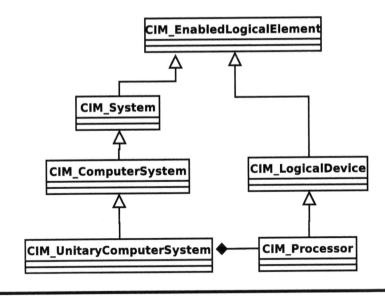

Figure 5.10   UML Representation of a Composition

Figure 5.11   Alternative UML Representation of a Composition

UML for Composition

There is a particularly strong type of aggregation: "composition." This is illustrated in Figure 5.10 where the filled diamond means that a CIM_UnitaryComputerSystem (a desktop, mobile, NetPC, server or other type of a single node computer) may include an aggregation of CIM_Processors; the processors exist within the CIM_UnitaryComputerSystem and are dependent on it, being created and destroyed when it is. A CIM_Processor, which incidentally *is a* CIM_LogicalDevice, really has no life outside of a CIM_UnitaryComputerSystem.

On some of the DMTF's drawings, a "diamond with a dot" notation is also sometimes used for composition; see Figure 5.11.

The *mof* Language

General

mof is a textual language used, like UML, to describe CIM models. It generally contains more details than the equivalent UML diagram, but is less easy for a human to read. Some tools are available to assist in the preparation of *mof* and I discuss these starting on page 273.

The *mof* language itself is defined in the Common Information Model (CIM) Specification (Document DSP0004 available from the DMTF Web site (http://www.dmtf.org). and I make no attempt here to describe the whole language. Instead I give a few pointers to the general syntax of the language and describe how:

- Instances are named: see page 66.
- Qualifiers are defined to specify the format and use of a property or class (e.g., this property has a maximum value of 42): page 69.
- Classes are defined: page 74.
- Association classes are defined: page 77.
- Indication classes are defined: page 78.
- Instances of classes are defined: page 79.

The best examples of *mof* code are probably the DMTF's core and common models themselves; you should download these from the DMTF Web site and study them. In general, the following rules apply to *mof* programs:

- Comments take the C++ and C forms:

```
// this is a comment

/* and so is this */
```

- Names are case insensitive (i.e. Disk, DISK and dISk all represent the same object, although it may not be sensible to rely on all implementations following this rule).
- The basic numeric datatypes are better defined than those of C/C++, being signed and unsigned 8, 16, 32 and 64-bit integers, and 4 and 8-byte floating point numbers. Additionally *mof* supports Booleans, strings, and datetimes as basic datatypes. These are straightforward with the possible exceptions of string and datetime, which I describe in Appendix D.

- There is a number of compiler directives supported through the use of the #pragma construction (a pragma is basically a way of expressing concepts outside the language). These include
  - #pragma namespace(). This pragma specifies the namespace of the following *mof* code. The concept of a namespace is discussed more fully starting on page 197.
  - #pragma locale(). This pragma specifies the language and country of the system. As I describe later, it is possible for some descriptive text in the *mof* to be specified in different languages. This pragma defines the actual locale of the management system and thereby allows a particular language to be selected. The format of the pragma is

    ```
    #pragma locale("ll_cc")
    ```

    where ll is a language code taken from ISO/IEC 639 and cc is a country code taken from ISO/IEC 3166. If no pragma locale is given, then it is assumed to be en_us: i.e. English as spoken in the United States of America. A useful explanation of this coding is given in RFC1766

    Having described this excellent structure for making CIM models international, I must confess that the DMTF core and common models are documented internally only in en_us. There is also a problem associated with locale—what if a French-speaking operator in Germany is managing a device installed in Croatia? What locale should be used?

    See page 145 for more discussion of locale and international character sets and languages.
  - #pragma include(). As with the C or C++ precompiler, this pragma simply allows another *mof* file to be included as though it were typed at that point.
- As with C++, constant strings can be continued over a line by closing the quotation marks and re-opening them at the beginning of the next line. Thus

```
"I met a "
"traveller from "
"an antique land"
```

is a single string constant comprising

```
"I met a traveller from an antique land"
```

- Enumerations (the equivalent of a C++ **enum**) are defined in two parts: a list of the keys, known as a ValueMap, and a list of the associated values, known as Values. This is illustrated in the following definition, taken from the DMTF's model of the class CIM_ProductProductDependency:

```
[Description (
        "The nature of the Product dependency. This "
        "property describes that the associated "
        "Product must be installed (value=2) or "
        "must be absent (value=3) in order for "
        "the Product to function."),
    ValueMap {"0", "1", "2", "3"},
    Values {"Unknown", "Other",
            "Product Must Be Installed",
            "Product Must Not Be Installed"} ]
    uint16 TypeOfDependency;
```

In this example, TypeOfDependency is an enumerated property which can take the values Unknown, Other, Product Must Be Installed, or Product Must Not Be Installed. These correspond to actual integer values 0, 1, 2, and 3. Note that this example also uses the string-continued-on-next-line convention described earlier.

The Naming of Parts

It must be possible to identify any particular instance of a class uniquely. In the domestic dog example illustrated in Figure 5.3, for example, we need to distinguish Fido (the dog) and Joe (the owner). To do this, we specify that some of Fido's properties (defined on his class or one of his superclasses) have the qualifier KEY. The combination of his namespace, class name, and the values of all keys forms a unique identifier for Fido.

Assume, for example, that the Name property in the *DomesticDog* class is Fido's only KEY. Then Fido might be identified informally as

```
root/acne:Dachshund.Name="Fido"
```

and Joe might be identified by

```
root/acne:DogOwner.Name="Joe Smith"
```

In each of these cases, the word "root/acne" is a namespace (presumably registered by the Acne Manufacturing Company for its components). By the way, although this name contains a slash and a slash is often used in operating systems to represent some form of hierarchy, this namespace is not hierarchical—it consists of one word containing the symbols r, o, o, t, /, a, c, n, and e.

If you are familiar with C++, then you already understand CIM namespaces. If not, then consider a namespace to be a domain (or container) within which names are unique. Thus you may have a class called *Dachshund* referring to a new car being developed and I may have a class called *Dachshund* referring to the type of dog, but as long as they exist in different namespaces, they remain unrelated.

The namespace is followed by the class name and then by a list of key values identifying the particular instance.

More formally, objects (either classes or instances of classes) are uniquely identified by a name known as an Object Path or Object Name. This name is effectively equivalent to a URL and has the following structure:

```
<namespacePath>:<modelPath>
```

where the two components also have structure as follows:

- Namespace Path
  - Namespace Type. This specifies the access protocol or API set that must be used to access the instance and any address necessary for reaching the instance. An example of a Namespace Type might be http.‖
  - Namespace Handle. This specifies the namespace within which the instance has been created.
  `http://47.2.34.2/root/cimv2` is an example of a Namespace Path.
- Model Path. This part of the name identifies the particular instance within the namespace. It takes the form

```
<class>.<key>=<value>,....,<key>=<value>
```

‖ Hypertext Transfer Protocol: a generic, stateless, object-oriented protocol described in RFC1945.

An example of a Model Path is[**]

```
CIM_BGPCluster.ClusterID=74
```

The full name of a CIM_BGPCluster instance might therefore be

```
http://47.2.34.2/root/cimv2:CIM_BGPCluster.ClusterID=74
```

This is the full name (Object Path) of an instance. The same format is used to identify a class but the "key=value" clauses are naturally omitted:

```
http://47.2.34.2/root/cimv2:CIM_BGPCluster
```

Weak Associations

I deferred discussion of weak associations until this point because you needed more background information. You are now equipped to dive into the following description.

A weak association is used to name an instance of one class in the context of an instance of another class. For example, the user *cwlh* may be logged on to several computers at the same time. A particular instance could always be named uniquely by including the particular computer as a property of the user, but this could mean unnecessarily duplicating a significant amount of information. Because there is likely to be an association between the user and the computer already in existence, it would be better to name the user by a combination of the user name (e.g., *cwlh*) and the computer name (e.g., *mercury*), provided that there is a weak association between the two.

A more realistic example of a weak association is given in Figure 5.12 where a CIM_MPLSService[††] is weakly associated with a router, modelled as a CIM_ComputerSystem. The full name of the CIM_MPLS-Service is

---

[**] BGP is a protocol used by two IP routers owned by different companies to exchange reachability information—see Glossary.

[††] MPLS: Multi-Protocol Label Switching: A layer 2.5 connection-oriented communications protocol—see Glossary.

```
CIM_MPLSService.CreationClassName="CIM_MPLSService",
        Name="MPLSService12",
        SystemCreationClassName="CIM_ComputerSystem",
        SystemName="Router6635"
```

As you can see, a CIM_MPLSService has four keys, all inherited from CIM_Service:

1. CreationClassName: The name of the subclass of CIM_Service to which this instance belongs.
2. Name: The name of this actual MPLS service.
3. SystemCreationClassName: The name of the subclass of CIM_System associated with this MPLS service. The System-CreationClassName property has a qualifier to say that it is propagated from the CIM_System class property CreationClass-Name.
4. SystemName: The name of the instance of the CIM_System associated with MPLS service. Again, this property is propagated from CIM_System.

Specifying Qualifiers

Qualifiers give additional information about classes, associations, properties, etc. They may specify the maximum length of a string property, the maximum value of a numeric property, whether a particular property may be modified (written) or only read. The CIM specification defines many qualifiers, some of which I explain here but you can also define your own for conditions specific to your system: "this property must be an IP address."

There is some confusion about whether an instance may have qualifiers. This could mean, for instance, that whereas a string property could be defined at the class level to have a maximum length of 10 characters, a particular instance of the class could have a qualifier saying that the maximum length of the string was only 8 characters. Whether or not this is legal is somewhat ambiguous at the moment and is being resolved in the DMTF—probably by making instance qualifiers illegal.

Some of the more important qualifiers predefined by the CIM standard include:

■ WRITE to indicate that a property may be changed by an operator.

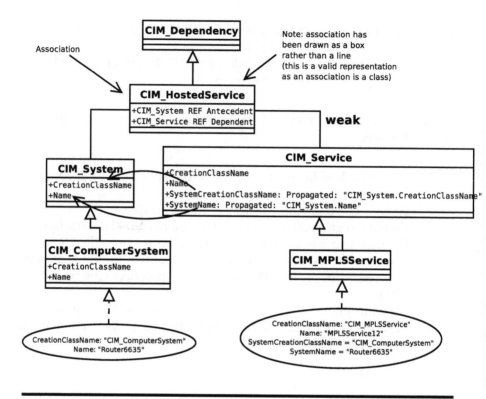

Figure 5.12   A Weak Association

- READ to indicate that a property may be accessed by an operator.
- KEY to indicate that the value of a property (or reference) forms part or all of the key of an instance, allowing the instance to be uniquely identified as described above. Thus, for example, in the definition of a CIM_BGPCluster in the network model, the ClusterID property is defined with the qualifier KEY as follows:

```
[Key, Description (
    "If a cluster has more than one route reflector, "
    "all of the route reflectors in the cluster need "
    "to be configured with a 4-byte cluster ID. This "
    "allows route reflectors to recognize updates "
    "from other route reflectors in the same cluster.")
    ]
uint32 ClusterID;
```

This means that a particular instance of a CIM_BGPCluster is identified by having a unique ClusterID. In general a class can have several keys and it is the combination of key values that makes each instance unique.

■ ABSTRACT to indicate that a class has been defined only to be a superclass for other classes. It is not possible to create an instance of a class which has the ABSTRACT qualifier. If you are familiar with the C++ language, this terminology will not be new—a C++ class which can only act as a superclass for other classes is called an abstract class. Sometimes italic text is used for the class name on the UML diagram to indicate that the class is abstract.

■ TERMINAL to indicate almost the opposite of ABSTRACT—this class may *not* be subclassed. Naturally, a class may not be both ABSTRACT and TERMINAL.

■ MAXVALUE to define the maximum value a property, method, or parameter may have.

■ REQUIRED to indicate that a property must have a genuine (non-NULL) value.

■ DESCRIPTION to allow a full textual description of a class, property, association, or parameter to be included in the *mof* in such a way that it is stored in the repository and made available at run time for enquiries. See Figure 5.14 on page 77 for a good example of the use of the DESCRIPTION qualifier.

■ VERSION to allow a version number to be associated with every class, association, or indication. As with the DESCRIPTION qualifier (or, indeed, any qualifier), you can read the value associated with the VERSION qualifier from the repository with a query. See Figure 5.14 on page 77 for an example of the VERSION qualifier.

■ PROPAGATED to allow a key property in one instance to be associated through a so-called weak association (see page 68) with a key property on another instance, allowing the number of keys to be reduced. I give an example of the use of the PROPAGATED qualifier on page 68.

■ DEPRECATED to discourage the use of a class. This qualifier is used in the DMTF's models where a class has been superseded by a different class, but the old class has been retained for backwards compatibility—the qualifier effectively means that you should not use the class for new models, but that it will remain for use in your existing models. Whenever the qualifier is used, it should point to the new, nondeprecated class.

■ MAPPINGSTRINGS to allow a particular property (or class) to be defined as equivalent to a particular property in another model. This is a particularly important concept during migration from a legacy management system to WBEM, and I give a slightly extended description in Appendix E.

- MIN and MAX to specify the way in which an association may or must be used. These qualifiers can be confusing because they appear at first glance to refer to the "wrong end" of the association and a few examples may be in order. Assume that an association links an instance of class A (antecedent) with an instance of class B (dependent). The following conditions may occur depending on the application:

  - Any number (including zero) of B instances may be associated with an instance of A. In this case no MIN and MAX qualifiers are needed.
  - At least one instance of an A is required for each instance of a B. This is represented by *mof* of the form
    ```
    [Min(1)]
      A REF antecedent;

      B REF dependent;
    ```
  - At most six instances of an A are permitted for each instance of a B. This is represented by *mof* of the form
    ```
    [Max(6)]
      A REF antecedent;

      B REF dependent;
    ```
  - Exactly one, two, or three instances of an A are permitted for each instance of a B. This is represented by *mof* of the form
    ```
    [Max(3),Min(1)]
      A REF antecedent;

      B REF dependent;
    ```
    MAX and MIN qualifiers are illustrated in Figure 6.3 on page 98, where they are used to specify that precisely one (not zero, not two) instance of CIM_ManagedElement must be associated with each instance of CIM_StatisticalData. This means that whenever an instance of CIM_StatisticalData is created, an instance of CIM_ElementStatisticalData must be created to associate it with exactly one CIM_ManagedElement.

These are a very small selection from a wide range of predefined qualifiers. You may also specify your own qualifiers. Assume, for example, that you require a qualifier on properties which requires that the property have a genuine (non-NULL) value on Tuesdays. This could be declared in the *mof* as follows:

```
qualifier RequiredOnTuesdays : Boolean = false,
                               scope (property),
                               flavor (DisableOverride);
```

(note the en_us spelling of "flavour"). This declaration defines a new qualifier called `RequiredOnTuesdays`, which can take the values **true** and **false** (i.e., it is a Boolean) and defaults to the value **false**. It can be applied to a property (but not, for example, to a class or association) and it may not be overridden in a subclass.

As the `RequiredOnTuesdays` example shows, qualifiers may have flavours. A flavour typically specifies whether a qualifier on a class should be inherited by subclasses and, if so, whether the subclasses may override it. In the `RequiredOnTuesdays` example, the qualifier is inherited by the properties in all subclasses of the class to which it is applied and may not be overridden there.

Be aware that if you do create your own qualifiers, then there is the possibility that the name you choose (`RequiredOnTuesdays` in this example) will clash with a name chosen in the future by the DMTF. The CIM specification includes a complete list of the current qualifier names and some names reserved for future use—obviously, you should avoid these, but it would probably be safer if you prepended your company's registered name to any qualifier names you choose (ACNERequiredOnTuesdays).

Another important flavor of a qualifier is **Translatable**. Consider, for example, the definition of the qualifier **Description** as given in the standard *mof*:

```
Qualifier Description : string = null,
    Scope(any),
    Flavor(Translatable);
```

This defines the qualifier **Description** as being a string which may be applied to any *mof* entity: property, class, etc. A description may also be translated into different languages, identified in the class or property definition by a pseudo-qualifier of the form `Description_ll_cc` where `ll` is the language code (from ISO/IEC 639) and `cc` the country code (from ISO/IEC 3166) of the language. Consider, for example, the dog-owning hierarchy illustrated in Figure 5.3 on page 54. The description of a (male) dog owner might be

```
Description_en_gb ("Chap who has a dog"),
Description_en_us ("Guy who has a dog"),
Description_fr_fr ("Mec qui a un chien"),
Description_fr_ca ("Gars qui a un chien"),
Description_de_de ("Bursche, der einen Hund hat"),
Description_de_at ("Mannaz, der einen Hund hat")
```

Another qualifier which has the **Translatable** flavor is **Values**: see page 66. This allows the Values associated with a value map to be written in several languages:

```
Valuemap {"1", "2", "3"},
Values_en_gb {"Dog", "Elephant", "Cat" },
Values_fr_fr {"Chien", "Elephant", "Chat"},
Values_de_de {"Hund", "Elefant", "Katze" }
```

For more details of some of the more nonintuitive standard qualifiers, see page 189. That section covers, in particular, the **OVERRIDE** qualifier which can be awkward to use correctly.

Specifying Classes

Apart from the inclusion of qualifiers in square brackets, the syntax for specifying a class in *mof* will be familiar to any C++ programmer:

```
[ <Qualifiers> ]
class <class name> : <superclass name>
   {
   <property or method>;
   <property or method>;
   .....
   <property or method>;
   };
```

Each property has the format:

```
[ <qualifiers> ]
<type> <name> { = <default> };
```

and each method has the format:

```
[ <qualifiers> ]
<return type> <method name>(<parameter>, ... <parameter>);
```

where, again, a parameter consists of a list of qualifiers in square brackets, followed by a type, followed by the parameter name. A simple property and method might therefore be coded in *mof* as follows (both taken from the CIM_LogicalDevice class):

```
[Deprecated { "CIM_PoweredStatisticalData.TotalPowerOnHours"},
   Description (
     "The total number of hours that this Device has been "
     "powered."),
   Units ( "Hours"), Counter ]
uint64 TotalPowerOnHours;


[Description (
     "Requests a reset of the LogicalDevice.  The return value "
     "should be 0 if the request was successfully executed, 1 "
     "if the request is not supported and some other value if "
     "an error occurred.  In a subclass, the set of possible "
     "return codes could be specified, using a ValueMap "
     "qualifier on the method. The strings to which the "
     "ValueMap contents are 'translated' may also be specified "
     "in the subclass as a Values array qualifier.") ]
uint32 Reset();
```

In addition to being able to specify a single property, *mof* can also specify that a property is an array of any of the basic types—integers, strings, Booleans, dates, and floating point numbers—but not of references (i.e., pointers to other objects). The syntax is again very similar to C/C++: square brackets after the property name, possibly enclosing a constant which specifies the array size. Thus Figure 5.13 defines an array of uint16s called LabelStack as one of the properties of the class CIM_MPLSCrossConnect, which *is a* CIM_Service. Note the use of the qualifier **Ordered** in this definition; *mof* allows three types of array to be specified:

1. A bag—an unordered array which allows duplicate entries. The array index has no significance for the items in a bag other than a guarantee that if all the possible indices are accessed then all items in the bag will have been accessed. There is no concept of ordinality for the items in a bag. If no qualifier is given, then an array is assumed to be a bag.

```
class CIM_MPLSCrossConnect : CIM_Service
    {
    ..... snip .....

    [Description (
        "Identifies a stack of labels to be pushed beneath "
        "the top label. Note that the top label identified "
        "in an instance of OutSegment ensures that all the "
        "components of a multipoint-to-point connection "
        "have the same outgoing label. This array is "
        "'Ordered' to maintain the sequence of entries."),
        ArrayType ("Ordered") ]
    uint16 LabelStack[];

    ..... snip .....
    };
```

Figure 5.13   An Example of an Ordered List

2. An ordered list (as in Figure 5.13)—like a bag, an ordered list allows entries to be duplicated, but the index now specifies an order for the entries when they are accessed.
3. An indexed array—like an Ordered List but where items can be overwritten but not deleted. The index starts at 0 and has no gaps.

An example of a class definition, taken from the Network Common Model, is given by Figure 5.14. This defines the class CIM_BGPCluster as a subclass of CIM_CollectionOfMSEs. Notice the following points about this definition:

■ The extensive use of the **DESCRIPTION** qualifier. This qualifier allows a textual description of any type of object (class, method, property, etc.) to be included in the repository. Because this makes it accessible for querying, using this qualifier is preferable to using comments in the *mof* code.
■ ClusterID is a key for a CIM_BGPCluster.
■ The version number of the class is included as a qualifier: again allowing it to be accessed by an operator.

```
// ===========================================================
// BGPCluster
// ===========================================================
    [Version ("2.6.0"), Description (
        "The BGP speakers in an AS are required to be "
        "fully meshed. This can lead to a huge number "
        "of TCP connections per router. One way to reduce "
        "the peering requirements is to use a route "
        "reflector. This is based on specifying one or "
        "more routers to act as focal points for IBGP "
        "sessions.\n\n"
        "The route reflector as a whole is called a cluster. "
        "It is logically divided into three types of routers: "
        "reflectors, clients of the route reflector(s), and "
        "non-clients of the route reflector. There can be "
        "more than one route reflector in a cluster, and "
        "there can be more than one cluster in an AS.") ]
class CIM_BGPCluster : CIM_CollectionOfMSEs
    {
        [Key, Description (
        "If a cluster has more than one route reflector, "
        "all of the route reflectors in the cluster need "
        "to be configured with a 4-byte cluster ID. This "
        "allows route reflectors to recognize updates from "
        "other route reflectors in the same cluster.") ]
        uint32 ClusterID;
    };
```

Figure 5.14   An Example Class Definition

Specifying Associations

Syntactically, associations look much like classes (not unreasonably—they *are* classes). A class becomes an association by the inclusion of the qualifier **ASSOCIATION** as illustrated in Figure 5.15.

The following points can be noted in this example:

■ The class CIM_RangesOfConfiguration is a subclass of the class CIM_Dependency from which it inherits and overrides the properties **Antecedent** and **Dependent**.

■ The keyword **REF** is used to define a "pointer" to the objects linked by this association. Note that all REFs must be keys.

```
// ============================================================
// RangesOfConfiguration
// ============================================================
    [Association, Experimental, Version ("2.7.0"), Description
      ( "This association connects address ranges to the "
      "OSPF area "
      "......much description cut for simplicity.......... "
      "to allow or disallow advertisements in the range.") ]
class CIM_RangesOfConfiguration : CIM_Dependency
    {
        [Override ("Antecedent"),
         Description (
         "The address range that is in the OSPF area "
         "configuration.") ]
    CIM_RangeOfIPAddresses REF Antecedent;

        [Override ("Dependent"),
         Description (
         "The OSPF area config that contains the range.") ]
    CIM_OSPFAreaConfiguration REF Dependent;

        [Description (
         "The address range is advertised (TRUE) or not "
         "(FALSE), see C.2 in RFC 2328.") ]
    boolean EnableAdvertise;
    };
```

Figure 5.15   An Example Association Definition

■ This association contains a property, **EnableAdvertise**, which is not a reference to another object. This is an example of an association with an additional property.

In the example shown in Figure 5.15 there are only two REFs—i.e., two objects are being associated. This is normally the case but, in principle, an association can relate any number (greater than 1) of objects.

Specifying Indications

Indications are just classes which inherit, directly or indirectly, from CIM_Indication. They have no special format.

Specifying Instances

In addition to defining classes, the *mof* language can be used to specify
instances of classes. This feature is used for instances where, for some
reason (e.g. the instance is completely static) the instance does not
depend on the run-time environment; it can be specified in advance. As
the *mof* is compiled, any providers handling the creation of instances
will be invoked as if the instance had been created by an operator. The
format is as follows:

```
instance of <class name>
    {
    [ <property qualifiers> ]
    <property name> = <value>;
    .....
    [ <property qualifiers> ]
    <property name> = <value>;
    };
```

For example, an instance of a CIM_BGPCluster as defined in Figure
5.14 could be specified in *mof* as follows:

```
instance of CIM_BGPCluster
    {
    clusterID = 73;
    };
```

To create an instance of an association, you must be able to specify the
instances which it associates. An instance of CIM_RangesOfConfigur-
ation as defined in Figure 5.15 could be coded as follows:

```
// definition of antecedent and dependent
// earlier in the mof

instance of CIM_RangeOfIPAddresses as $LHS
                        { ... definition ... };
instance of CIM_OSPFAreaConfiguration as $RHS
                        { ... definition ... };

// definition of the association instance

instance of CIM_RangesOfConfiguration
```

```
{
Antecedent = $LHS;
Dependent = $RHS;
EnableAdvertise = true;
};
```

Note that, in this case, the Antecedent and Dependent have been set to point to other instances created earlier in the file by using the technique known as aliasing—specifying an alias for an instance when it is specified so that it can be referenced elsewhere (forward references and even circular references are allowed). Without this technique, it would have been necessary to specify the full Object Path of the instance (see page 64 above).

Summary

Models form abstractions useful to different operators managing a device or service. These models abstract away the actual hardware and software implementations and present a common picture to operators, independent of the particular device or service implementation. In CIM such models may be expressed graphically in UML or textually in *mof*.

Frequently Asked Questions

FAQ 10 *Various qualifiers exist to allow me to specify, for example, that a particular numerical property has a maximum value of 10 or a string property has a maximum length of 23 characters. How can I encode constraints between properties, particularly properties in different classes (for example, this numeric property has a maximum value of 72 unless property X in class B has a value of "F" in which case the maximum value of this property is 97)?*

This cannot be done except by making an informal statement in the description of the property. It would, of course, be possible to use a formal definition of your own devising in the description and allow your own software to interpret it; for example:

```
[Description (
    "This property defines the size "
    "--->maxvalue dependsOn(B.X == 'F' ? 72 : 97;"
)]
```

Here I have assumed that you have defined your own escape sequence
(--->) and your own constraint language based on a sort of C++ syntax.

Descriptions are used informally at many points in the CIM schema.
One of the properties, for example, of the CIM_EthernetPort class is
defined as follows:

```
[Description (
  "Capabilities of the EthernetPort.  For example, the "
  "Device may support AlertOnLan, WakeOnLan, Load "
  "Balancing and/or FailOver.  If failover or load "
  "balancing capabilities are listed, a SpareGroup "
  "(failover) or ExtraCapacityGroup (load balancing) "
  "should also be defined to completely describe the "
  "capability."),
    ValueMap {"0", "1", "2", "3", "4", "5"},
    Values {"Unknown", "Other", "AlertOnLan", "WakeOnLan",
        "FailOver", "LoadBalancing"},
    ArrayType ("Indexed"),
    ModelCorrespondence {
"CIM_EthernetPort.CapabilityDescriptions"} ]
uint16 Capabilities[];
```

The informal description places a constraint on this property—if Capabilities contains Failover, then an instance of the CIM_SpareGroup class
should also exist. There is currently no way to express this formally,
but there is a movement within the DMTF to add a more powerful
constraint language of this type to *mof*.

Note, however, that the presence of a qualifier stating that the maximum value of a particular numerical property is 56 does not mean that
most WBEM servers will enforce such a condition. The qualifier is
there to be read by providers and clients which are expected to enforce
the constraint.

FAQ 11 *Can an association associate two associations?*

I assume this question means, "Can an association relate one association with another association?" If so, the answer is "yes." This is clear from the DMTF's metamodel where it is stressed that:

1. An association is a normal class.
2. An association can connect any two classes.

FAQ 12 *Can an association associate classes or instances in different namespaces or on different hosts?*

Yes. The full object path is included within each reference.

FAQ 13 *Can an association associate more than two classes?*

Again the answer is "yes." The Common Information Model Specification (DSP0004), version 2.2, is very clear about this:

> *The number of references in an association class defines the arity of the association. An association containing two references is a binary association. An association containing three references is a ternary association. Unary associations (associations containing one reference) are not meaningful... When an association is subclassed its arity cannot change.*

Having given the answer which is correct by the specification, I ought perhaps to give some advice: do not use more than two references in an association unless there is no other possibility because many WBEM server implementations do not handle more than two references properly.

Often, you can avoid using a four-way association between classes A, B, C, and D by making an association, called assoc1, between class A and class B, and another association, assoc2, between class C and class D. You can then establish an association between assoc1 and assoc2 and end up with three binary associations rather than one four-way one. The choice is yours: perhaps the four-way association is more meaningful in your model, perhaps the three two-way ones are.

FAQ 14 *I have searched the DMTF's UML diagrams that I downloaded from the Web for some of the classes you mention (e.g., CIM_RangesOfConfiguration) but cannot find them. Why?*

Probably because you are searching for the full name, including the CIM_ schema prefix. In their UML diagrams, the DMTF does not include the schema if it is CIM. Yes, I agree, this is really annoying and you will notice that, whenever I have drawn DMTF UML in this book, I have added the prefix. The DMTF does, of course, use the prefix in the *mof* code.

FAQ 15 *When coding a class definition in* mof, *how do I indicate that the superclass is in a different namespace?*

You cannot! When your class is defined by a *mof* statement such as

```
class myClass : yourClass { ......
```

and you pass it through a *mof* compiler for loading into the WBEM server, the WBEM server will expect to find **yourClass** already defined in the namespace. Note, however, that an instance of an association in one namespace may refer to objects in other namespaces.

FAQ 16 *Does the order of the keys matter in an Object Path?*

No. The following two object paths are identical:

```
http://47.2.3.2/root/cimv2:ACNE_C1,key1="fred",key2="joe"
http://47.2.3.2/root/cimv2:ACNE_C1,key2="joe",key1="fred"
```

FAQ 17 *What is the plural of "schema"?*

This is another source of friction within the modelling community. The correct plural is "schemata" in the same way that the plural of "lemma" is "lemmata." Many sources, however, use the abomination "schemas."

FAQ 18 *Why do associations not inherit from CIM_ManagedElement?*

Because an association can only be a subclass of another association. This means that two trees are necessary: one for classes which are not associations, rooted in CIM_ManagedElement, and one for associations, which have no root.

FAQ 19 *Within* mof *a class can only inherit from one other class. The concept of multiple inheritance, as in C++, is not supported. How can I work without multiple inheritance?*

By using associations. Inheritance is one way to move from one class to another, but a similar trick can be played (between instances) by using associations. There is one particular association, CIM_LogicalIdentity, which associates two CIM_ManagedElements (i.e., two anythings). It is designed to be used to indicate that two CIM_ManagedElements represent different aspects of the same underlying entity.

FAQ 20 *I understand that the WBEM server acts as a broker between the clients and providers, but how much checking does it do? Does it, for example, check the types and ranges of properties in an instance that a client is creating before it passes those parameters to the provider?*

The specification does not give direction on whether or not the WBEM server should check parameter values. Most implementations do no checking at all for performance reasons, arguing that the providers will almost certainly need to do additional range checking anyway before they act on the properties. This means that if you define a property as follows:

```
class myClass
    {
    [maxValue(10)]
    uint8    myValue;
    };
```

then a client could create an instance of this class with myValue set to 12 (or even, given that the value will be passed in a string rather than as a uint8, 15345) and the WBEM server would probably not notice. The argument is that the provider probably knows more about the actual allowed values of myValue (perhaps it is only allowed to be as large as 10 on Tuesdays, its maximum being 8 on all other days) and will be doing checking anyway—any additional checking in the WBEM server would adversely affect performance. Similarly, when an extrinsic method is invoked, the WBEM server is unlikely to perform any checking on the parameter types—WBEM is not CORBA.

FAQ 21 *Why does the* mof *language not contain support for feature X, which would make my modelling very much easier?*

This is a common question. Why, for example, is there no standard qualifier for specifying that a particular property holds an IP address? This would make *my* life much easier.

The DMTF is an open body. Convince other people of the worth of your suggestion and participate in the discussions at the DMTF to get your ideas adopted. You are only allowed to complain if you have worked hard to get your ideas adopted and everyone else is too stupid to agree with you.

Chapter 6

Standard Models

This section generally describes version 2.7 (final) of the standard models prepared by the DMTF. Where there have been major changes between version 2.7 and 2.8 (in particular, in the User Common Model) I have tried to use the 2.8 classes even though they were not firm at the time of writing. Although these models will be changed, backwards compatibility will be guaranteed for all 2.x releases. You should, however, check any references I make here against the most recent version of the models.

The Core and Common Models

The DMTF has co-ordinated the preparation of various models* which can be divided into three categories: a Meta-Model, a Core Model and numerous Common Models.

In addition to these three DMTF-produced models, it is expected that individual companies will extend these models to encompass features unique to their products. These models are known as extension schema and they are created by subclassing from the Core and Common models.

The Meta-Model describes the components which can be used to build models (e.g., an Indication *is a* Class which *is a*

---

* Freely available from the DMTF Web site.

NamedElement. A Method *is a* NamedElement and a Class aggregates a number of Methods.). Although this meta-model has great theoretical significance, it does not help you build a model of a particular system and I will not consider it further.

The Core Model which contains concepts applicable to the management of anything from my toaster to the largest telecommunications network.

You can think of the Core Model as the basic tree on which all other models are hung—it contains a relatively small set of classes, associations and properties giving a basic vocabulary for describing managed systems.

It is not anticipated that the Core Model will ever undergo substantial change because it is very generic and many Common Models (see below) have already been written which rely on it. I give more information about the Core Model on page 90.

The Common Models which are listed in Table 3.1. These describe the management structure of a number of particular applications: storage, networking, desktop computing, etc. I explain a few of these starting on page 98.

Versions of the Model

The DMTF has instituted strict rules regarding the way a model may change between versions and version numbers in the form `major.minor.update` (e.g., 2.7.1) have been added to all classes. Releases where the major release number does not change (e.g., between version 2.7.2 and 2.8.1) are known as "point releases." Releases where the major version number changes are known as "major releases."

For point releases, a number of backward compatibilities are guaranteed and a formal process is used to make changes to the model.

Assume it was decided that introducing a CIM_Toaster class into the device common model in a previous version had been a bad idea. Simply removing it might break programs already running in products and so it would first have the **DEPRECATED** qualifier attached to it, if possible pointing the programmer to the class which should be used instead. By deprecating the class in this way, its use is discouraged, but programs already using it continue to work.

Adding a new class to the model is a simpler problem because maintaining compatibility with earlier releases is not an issue. Although a version of the model is in flux and discussions are in progress within the DMTF, the new class will have the **EXPERIMENTAL** qualifier added to it. This discourages engineers from making use of a class which may disappear before the formal point release. Once the discussion is over and the point release is ready, the EXPERIMENTAL qualifiers disappear.

Major releases provide the DMTF with the opportunity to clean up all the deprecated classes and properties. Of course, this breaks backward compatibility by removing classes on which programs may be relying. To date there has been no major release and the precise mechanism is undecided and unproven.

The Logical/Physical Distinction

One concept of fundamental importance when considering the core and common models is the distinction between physical and logical entities. This may seem to be an obvious distinction but consider the question: "Is the network adaptor in my desktop computer physical or logical?"

You may answer "physical" without needing to take a great deal of thought—you may remember holding a network adaptor in your hand at some time. But the test to apply is, "Can you attach a label to it?" Certainly you can stick a label onto the printed circuit board which implements the network adaptor, but the network adaptor itself is a logical and abstract concept. What you have held in your hand is actually not a network adaptor but a printed circuit board which, among other things, implements a network adaptor. If the printed circuit board also implements a modem and a video interface, then the physical network adaptor is even harder to find.

The distinction which the DMTF makes in its Core Model white paper (DSP0111) is as follows:

> Physical Elements, occupying space and conforming to the elementary laws of Physics. The Physical Element class represents any component of a System that has a physical identity—it can be touched or seen.
>
> Logical Elements, representing abstractions used to manage, configure and co-ordinate aspects of the physical or software environment. Logical Elements typically represent Systems themselves, System components, System capabilities and software.
>
> The distinction between Logical and Physical Elements is fundamental to the structure of the Core Model. The principal distinguishing feature of a Physical Element is that it cannot have a "realization" (per Web-

ster's[†] definition of realization, it cannot be "brought into being"). It can be composed of parts, but there is no sense in which, for example, a system enclosure is "realized" (or "brought into being") by a piece of molded plastic; it simply *is* a piece of molded plastic.

Logical Elements (especially Logical Devices) can be "realized" by installing Physical Elements and/or software. For example, it is not possible to attach a label to a modem. It is only possible to attach a label to the Card that "realizes" the modem. The same card could also "realize" a LAN adapter. These tangible Managed System Elements have a physical manifestation of some sort. However, the physical manifestation is very different than its management aspects and attributes. The latter are addressed by the Logical Elements, realized from the physical, usually accessed via software.

The Core Model

Understanding the Core Model is a crucial part of building your own extension schema—unless you start with a blank sheet of paper and discard the available models in favour of "rolling your own," your components will be added somewhere beneath the Core Model. Getting your components correctly positioned is not trivial and may be difficult to adjust later.

The DMTF publishes a white paper, DSP0111, describing the Core Model that makes very useful reading.

Almost every class in CIM ultimately *is a* CIM_ManagedElement or, to put it another way, CIM_ManagedElement is an abstract superclass of every other CIM class (the principal exception being associations). CIM_ManagedElement provides three very general properties: Caption (a one-line description of the object), Description (a fuller textual description of the object) and ElementName (a simple name for the object).

Figure 6.1 illustrates a few of CIM_ManagedElement's subclasses which I discuss in the subsections below.

[†] A dictionary of American English.

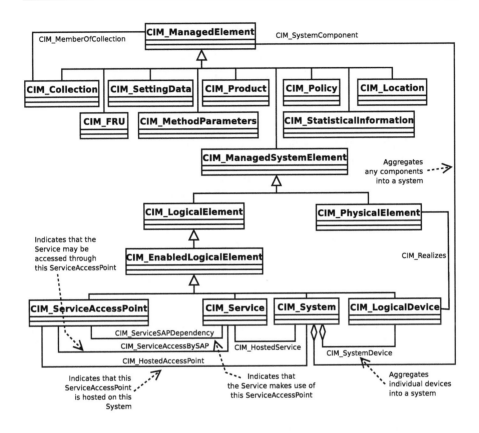

Figure 6.1   Part of the Core Model

CIM_ManagedSystemElement

This is the superclass of the great Physical/Logical split, its subclasses being CIM_PhysicalElement and CIM_LogicalElement.

You can consider every managed component of a system, whether software or hardware, to be a CIM_ManagedSystemElement. Thus, software modules, cards, integrated circuits, racks, frames, etc. are all CIM_ManagedSystemElements. The class defines properties for the date and time at which the element was installed, a name for the element and the element's state (Unknown, OK, Degraded, Stressed, Stopped, Dormant, etc.).

Note that, as defined in CIM_ManagedSystemElement, the Name property is not a key (i.e., cannot be used to identify a particular instance). It is anticipated, however, that you will derive subclasses from CIM_ManagedSystemElement (or, better, from one of its subclasses) and introduce the key property at a lower level.

CIM_ManagedSystemElement's Logical Sub-Tree

CIM_LogicalElement, the head of the logical tree beneath CIM_ManagedSystemElement, is very abstract and provides no new properties but, as seen in Figure 6.1, it acts as a superclass for two very important concepts, Systems and Services:

1. A **System** is a collection of elements working together to provide a particular functionality. Although the class CIM_System is defined in the core model, it is sufficiently important to be a common model in its own right. I describe it starting on page 99.

2. A **Service**, in the CIM sense, is a "function" of some sort which the device offers. Examples of a service might be voice mail and call forwarding on a PBX; a route calculation, source routing or forwarding in an IP router or user authentication, or printing on a computer. To determine the Services provided by a device, put yourself in the position of a user and ask yourself the question, "What does this device actually provide? Why would I want one?" The answer would not typically be "it provides a 10-GB disk drive and a POSIX API." More likely the end user would say, "it handles my voice mail and word processing." The Services then are voice mail and word processing.

   In order for a Service to be used, it needs at least one point through which it may be accessed: a Service Access Point. In order for the voice mail Service to be useful, it must provide a "way in" using, say, a telephone touchpad and display which I may or may not be authorised to use. I may also be able to access my voice mail from my computer—the Service then needs to model a further Service Access Point. Obviously CIM_ServiceAccessPoints need to be connected with their CIM_Service and this is achieved through the use of an association: CIM_ServiceAccessBySAP as shown in Figure 6.1.

   Remember that CIM is concerned not with the implementation of an entity but with its management. For example, CIM describes neither how a voice mail service is implemented nor how the service is accessed (the Application Program Interface, API). Instead it models the function provided by the service (using the class CIM_Service) and the way it may be accessed (using the class CIM_ServiceAccessPoint).

   So, my voice mail could be modelled as a CIM_Service and the different methods by which I may access it could be modelled as CIM_ServiceAccessPoints connected to the CIM_Service by means of associations. An operator using a WBEM client could then provision the voice mail service and enable or disable my access to it.

Note also that the voice mail service is not modelled as the computer which actually holds and distributes the voice mail or its associated software. The computer and software may well be CIM_LogicalDevices and these are managed independently of the service.

Having described a particular CIM_Service, it is essential to be able to link the service to the CIM_System which hosts it. This is done by using the CIM_HostedService association.

CIM_ManagedSystemElement's Physical Sub-Tree

CIM_PhysicalElement, the head of the physical tree, provides a large collection of additional properties including a tag (acting as a key), manufacturer, model number, stock-keeping code, serial number, version, part number, and date of manufacture.

You may be surprised to see a key this high in the class hierarchy and wonder why the physical element is not identified by reference to the device in which it is fitted. The tag (which is just a string and may contain an asset number or similar identifier) is in this position because of the potential mobility of a physical element. In general, a physical element (e.g., a card or removable disk) may be removed from a particular shelf or computer and be stored in a repair shop for some time before being inserted into a totally different system. Because a physical element has an independent existence, irrespective of the system into which it is fitted, it is given a unique key to allow it to be tracked as it moves about.

Because management protocols such as SNMP are primarily interested in the management of physical devices, it is not surprising that many of the properties of CIM_PhysicalElement map to SNMP MIB variables. PartNumber, for example, maps to MIB.IETF—Entity-MIB.entPhysicalModelName.

CIM_Product

If you are producing a model of a new product, it is possible that you will latch onto the CIM_Product as a means of defining your new product. While this is a possible use for CIM_Product, the class is more normally used to describe a complete bought-in product. Before subclassing from CIM_Product, ask yourself: "Did I purchase this entity as a single product, does it have a serial number, is there a legal agreement between me and the vendor who produced this product and do I have a support agreement for it?"

Unless the answers are generally "yes," then what you have may not be a product in CIM terms. A CIM_Product has properties such as a serial number, vendor's name, version number and warranty dates.

CIM_FRU

Informally, you can think of a field-replaceable unit (FRU) as the thing which a maintenance technician carries on her bicycle when going into the field to repair a fault. It may be a board containing processors, ports, memory, etc. It may be a power supply unit. Whatever it is, it can be unplugged and replaced.

Because it is a replaceable unit, the CIM definition of a CIM_FRU includes an identifying number (serial number, die number, etc.), the name of the supplier, a description of the FRU and a Boolean to indicate whether the customer is allowed to replace the unit or not.

From the modelling point of view, an FRU is a collection of physical elements, products and software features. Although this may seem to be a very computer-oriented view of an FRU, reflecting the history of the DMTF, it is a useful abstraction for most FRUs.

CIM_SettingData

Typically the state of a device is represented by two values: the state in which you, the operator, want it to be and the state in which it actually is. You may, for example, have configured a slot on router 32 on site X to contain a Gigabit Ethernet interface card (required state) but the field technician may have incorrectly inserted a 80-Baud Telex card (actual state).

The purpose of keeping both the required and actual state in this way is useful in preventing misleading information being given to an operator. Consider a configuration command which takes some time to complete: change the speed of a supertanker, for example. Assuming that the tanker is travelling at 20 knots, an operator might want to change the speed to 10 knots. If the desired and actual states were not separate then, while the tanker is slowing down, a period I imagine of some minutes, an operator could not tell what was actually happening. With two distinct values for the required and actual states, the operator would see (20,10), (19,10), (18,10), and finally (10,10) as the pairs of values.

Although both the required and the current states may be represented as properties on the instance of the managed element, it is also

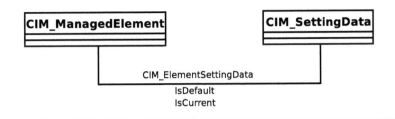

Figure 6.2   CIM_SettingData and CIM_ManagedElement

sometimes convenient to have the required state in a different class, associated with the managed element. This is the purpose of CIM_SettingData. This class was introduced in version 2.7 of the CIM core model to replace a rather idiosyncratic class known as CIM_Setting. CIM_Setting is not recommended for new designs, so I will not describe it here.

The purpose of an instance of CIM_SettingData is to hold a group of related and relatively static parameter values. For example, the values which define a computer monitor (vertical and horizontal refresh rates, horizontal resolution, interlaced or noninterlaced scan mode, etc.) would make good candidates to be drawn together in an instance of CIM_SettingData. They are clearly related because the whole group needs to be defined to make a monitor work and they are changed relatively infrequently. A counter which changes frequently would not be a suitable candidate for CIM_SettingData.

The association CIM_ElementSettingData between the CIM_SettingData class and a CIM_ManagedElement is illustrated in Figure 6.2.

CIM_SettingData itself has only two properties: an InstanceID and an ElementName. The InstanceID is particularly complex as it needs to be unique within a namespace; I discuss the format of this property starting on page 195.

CIM_Collection

A CIM_Collection, as its name implies, allows items to be grouped in some way. A straightforward example of this appears in the DMTF's User Model (see page 114). A user, typically a human being, can usefully be modelled as being a member of various groups: his or her department, the group of people carrying out the same role, etc. As illustrated in Figure 6.10 on page 115, these groups are modelled as CIM_Collections.

Of course, once a Collection class has been defined, you need a way of saying what items are members of it. This is achieved by using the CIM_MemberOfCollection association which, as you can see from Figure 6.1, allows any CIM_ManagedElement (i.e., anything!) to be associated with a CIM_Collection.

Collections are powerful when you want to associate something, not with instances of individual classes, but with instances of many classes. An example might be CIM_SettingData which is associated with classes A, B, C and D. When building the model without the concept of a Collection, you would have to build instances of associations between the SettingData and A, between the SettingData and B, etc., resulting in an explosion in the number of instances. This technique would also hide the fact that, in having the same SettingData, the four classes are actually related.

It would be better to create a collection, inheriting, perhaps indirectly, from CIM_Collection and drawing A, B, C, and D together. The collection could then represent whatever it is that they have in common and share the SettingData. The SettingData would then be associated with the Collection:

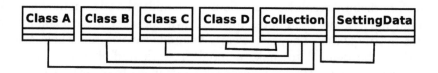

I give another example in Chapter 9 where I work through designing the model for a small device. Figure 9.9 on page 183, in particular, illustrates the collection hierarchy and an association created to link a trunk port with a number of telephony ports.

CIM_Location

Physical devices are assumed to have a particular location in space and, for many management purposes, the location is important. Rather than make the location a Property of a class such as CIM_PhysicalElement, the core model defines a CIM_Location class and an association, CIM_PhysicalElementLocation, to link a Physical Element with a Location. This reduces the amount of duplication (and the resulting difficulties in updating) if several Physical Elements are in one location.

Effectively, as you would expect, the CIM_Location class just defines an address. Its properties are:

- Name: A string which acts as a key and defines the location.
- PhysicalPosition: Another string which somehow defines the location. This could contain the latitude and longitude, or the site, building, rack, and slot number, depending on your application.
- Address: Yet another string designed to contain a street address.

CIM_StatisticalInformation and CIM_StatisticalData

Most devices create statistics, perhaps counts of calls or packets for billing purposes, perhaps information about packet queue lengths or module failure rate for debugging or performance management purposes. The CIM_StatisticalInformation and CIM_StatisticalData classes are designed to hold this type of information. Of the two, CIM_StatisticalData is the younger, being introduced in version 2.7 of the model, and although CIM_StatisticalInformation is not officially deprecated, I would recommend using CIM_StatisticalData.

A possible question, of course, is why classes such as these are necessary. Because the statistics in question will relate to a particular Managed Element, why should the counters, etc., not be made properties of those Managed Elements? The answer lies to some extent in the nature of statistical counters—they are typically dynamic and have to be fetched from devices when required and often require some form of computation to manipulate them before they are ready to be displayed to an operator. If they were properties of the Managed Element, then every time an operator totally uninterested in statistics requested information about an instance of a Managed Element, those values would have to be retrieved and calculated, placing a possibly heavy burden on the system.

So, instead of including the statistics as properties on the Managed Element, an instance of CIM_StatisticalData is created and linked to a particular CIM_ManagedElement by means of the CIM_ElementStatisticalData association. The full *mof* definition of this association is given in Figure 6.3 as it neatly illustrates the use of the **Max** and **Min** qualifiers described on page 72.

A good example of the use of CIM_StatisticalData is described in the paper "Generic On-Line Discovery Of Quantitative Models For Service Level Management" by Diao *et al.* of the IBM T.J. Watson Research Center, presented at the 2003 IEEE Conference on Integrated Management. This team was interested not in CIM *per se*, but in determining which of the 500 or so metrics available from their database system were actually significant in estimating management performance. The team, therefore, was a user rather than a designer of a CIM management sys-

```
[Association, Version ("2.7.0"), Description (
    "CIM_ElementStatisticalData is an association that relates "
    "a ManagedElement to its StatisticalData. Note that the "
    "cardinality of the ManagedElement reference is Min(1), "
    "Max(1). This cardinality mandates the instantiation of "
    "the ElementStatisticalData association for the "
    "referenced instance of CIM_StatisticalData. "
    "ElementStatisticalData describes the existence "
    "requirements and context for the CIM_StatisticalData, "
    "relative to a specific ManagedElement.") ]
class CIM_ElementStatisticalData
    {
    [Key, Min(1), Max(1), Description (
        "The ManagedElement for which statistical or metric "
        "data is defined.") ]
    CIM_ManagedElement REF ManagedElement;

    [Key, Description (
        "The statistic information/object.") ]
    CIM_StatisticalData REF Stats;
    };
```

Figure 6.3 The CIM_ElementStatisticalData Association

tem and their use of the information collected in CIM_StatisticalData instances illuminates a user's perspective.

The CIM_StatisticalData class itself defines a number of properties, including:

- An InstanceID which must be unique within the namespace. As for the CIM_SettingData class, see my description on page 195.
- An ElementName which is a user-friendly name for the instance.
- A method which can be invoked to reset one or more of the counters contained in the statistical data.

The Common Models

The CIM common models are listed in Table 3.1 on page 20. I describe the system, network, user, and policy models later because these are of particular importance to the telecommunications industry. Additionally, the Interop and Event models are described elsewhere in this book as indicated in the table.

The System Model

A **System** is a collection of elements which work together to provide a particular functionality. To some extent a system is therefore just a collection but, to be a system, the collected items must be synergistic— the collection must offer more than the sum of its components.

Typically a system will have a name and this is another way of recognising a system. Except, for example, in very specialised applications, a computer would have a name (e.g., saturn@hobbs) whereas its keyboard would not. The computer might therefore be a System, but the keyboard would not. Within the System Common Model, CIM_ComputerSystem *is a* CIM_System and its presence this high in the class hierarchy may depend more on the history of the CIM models (arising from the desktop computing environment) rather than its actual importance. Having said that, computing devices do appear in many devices being managed. Remember, however, that CIM_System *is a* CIM_LogicalElement, not a CIM_PhysicalElement—we are talking here of the logical function of computing, not the physical computer.

Not all Systems are as obviously Systems as is CIM_Computer- System. An OSPF Area (if necessary, see the glossary entry for OSPF), represented by CIM_OSPFArea in the Network Common Model, is also, ultimately, a CIM_System through being a CIM_RoutingProtocolDomain which is, in turn, a CIM_AdminDomain.

So, does an OSPF Area comply with the guidelines for being a System? Of what is it a collection? Does it add any synergy not provided by the individual elements? Could you name it?

OSPF is defined in the IETF specification RFC2328, which contains the statement:

> OSPF allows collections of contiguous networks and hosts to be grouped together. Such a group, together with the routers having interfaces to any one of the included networks, is called an area.

Given this description, an OSPF Area is certainly a collection—the RFC even uses the word "collection" directly. RFC2328 goes on to say:

> Each area runs a separate copy of the basic link-state routing algorithm. This means that each area has its own link-state database and corresponding graph.

which implies that an OSPF Area is not just a random collection of unrelated entities—by grouping them into a collection like this extra functionality emerges: routing tables based on common link-state databases can be calculated and IP packets can be efficiently routed.

Of course, an OSPF Area can also be named: it has an Area Identity defined in the *mof* file as follows:

```
[Description (
     "The area ID of this OSPF area, see C.2 "
         "in RFC 2328."),
     MappingStrings
         {"MIB.IETF|OSPF-MIB.ospfAreaId"} ]
   uint32 AreaID;
```

These are the characteristics of a CIM_System and it is therefore right that a CIM_OSPFArea should be a CIM_System.

Summary of the System Model

A system is a collection which is more than the sum of its parts.

The Network Model

In other parts of this book I have given brief explanations of network terms (OSPF, BGP, etc.) as they have arisen. I assume that, if you're reading this chapter, then you're probably interested in, and knowledgeable about, telecommunications networks and so I will drop the explanations.

The Network Common Model is enormous, is incomplete and has undergone radical change as the DMTF's Networks Working Group has partially unwound and rebuilt the original models—work which is still in progress. In version 2.7 of the core and common files, the Network

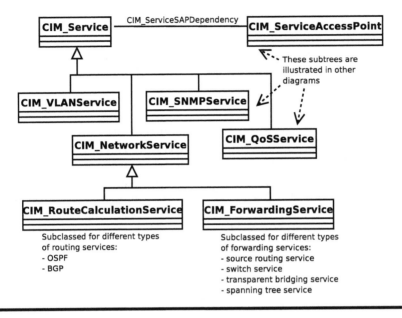

Figure 6.4   Part of the Network Model

Model accounts for 11,422 of the 52,218 total lines of *mof*—i.e., 22 percent of the core and common *mof* is Network Model. Given this size it is difficult to find a way into the model and understand its basic structure. To help with this, the DMTF has issued a white paper, DSP0152, dedicated solely to the Network Model (also one dedicated to the OSPF submodel: DSP0160).

The best way into the Network Model is through its services, but before diving into the details, it is perhaps useful to discuss the use of the term "Service" in the Network Model: the internal protocols used by the IP Network being managed are modelled as services (for example, CIM_OSPFService *is a* CIM_RouteCalculationService which *is a* CIM_NetworkService which *is a* CIM_Service). Generally, OSPF would not be seen by users of a network but, as I have said previously, services can exist at different levels and this is one example of that flexibility.

As illustrated in Figure 6.4, the fundamental services which it is assumed that the network offers are:

■ Virtual Local Area Network (VLAN) Service
■ So-called Network Services; the forwarding of packets at layer-2 or layer-3 and calculation of routing tables

- SNMP (yes, one management system managing the service offered by another!)
- Quality of Service (yes, Quality of Service (QoS) is a service—if you find this difficult to understand, then wait a page or so for the explanation.)

These services clearly illustrate the current orientation of the Network Model—towards enterprise IP networks running on Local Area Networks. Within the full Network Model, concession *is* made to the Wide-Area Network; Local Area Networks may be modelled as being connected by CIM_NetworkPipes: very simple representations of the underlying Wide Area Network connections. The details of Wide Area Networks (SDH, SONET, Wavelength Division Multiplexing, fibre management, etc.) and layer 2 networks (MPLS,‡ ATM, Frame Relay) are currently missing from the models. The omission of the WAN and layer 2 networks is likely to be remedied soon.

The Network, SNMP and QoS Services are sufficiently important to be considered in detail; see the following subsections.

Network Service

As I described on page 92, services are always associated with CIM_ServiceAccessPoints. So, if routing and forwarding are to be services, we expect to see their Service Access Points: the places at which these services "appear" and can be accessed. To this end, CIM_ServiceAccessPoint is subclassed to CIM_ProtocolEndpoint as illustrated in Figure 6.5. The endpoints of different protocols (IP, BGP, OSPF, TCP, UDP) are all represented as CIM_ProtocolEndpoints.

We may want to associate protocol endpoints in one of two ways:

1. By saying that endpoint A is (or may be) connected to endpoint B. Clearly, for this to occur the two endpoints must be of the same type: connecting a LAN Endpoint to a TCP Endpoint is meaningless. The association CIM_ActiveConnection which is defined between two endpoints (actually between two CIM_ServiceAccessPoints and inherited by the endpoints) is used to indicate this type of connection.

2. By saying that endpoint A "rests on" endpoint B. This occurs in multilayer networks: BGP endpoint A runs on TCP endpoint B, which runs on IP endpoint C, which runs on Frame Relay endpoint D, which runs on ATM endpoint E, which runs on

---

‡ OK, MPLS is really layer 2.5.

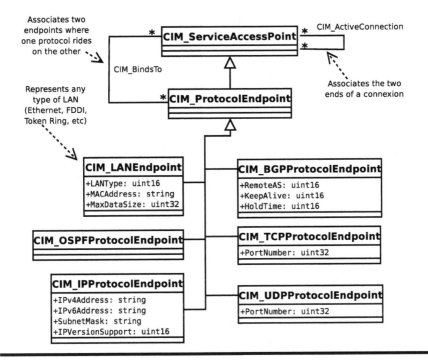

Figure 6.5   Part of the Protocol Endpoint Model

DWDM endpoint F, which runs on fibre endpoint G. Endpoints A to G are all in the same place, but are at different protocol levels. This type of relationship between endpoints is indicated by the CIM_BindsTo association.

The only concessions made in the Network Model at present to the underlying Wide Area Network are the classes describing a so-called "Network Pipe." Reasonably enough, the concept of a Pipe is taken from an ITU-T specification: M.3100 "Generic Network Information Model." To bind these two specifications together, the *mof* definition of the CIM_NetworkPipe model contains the mappings:

```
MappingStrings {"Recommendation.ITU|M3100.Pipe",
    "Recommendation.ITU|M3100.TrailR1",
    "Recommendation.ITU|M3100.ConnectionR1",
    "Recommendation.ITU|M3100.SubNetworkConnection"}
```

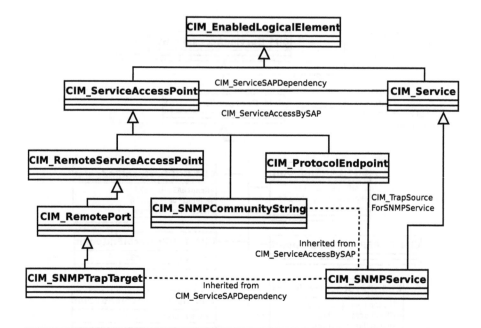

Figure 6.6   The SNMP Sub-Tree

This is an example of the mapping string used to map a CIM class to an ITU rather than SNMP definition: for a more complete description of the **MappingStrings** qualifier, see Appendix E.

The CIM_NetworkPipe class contains a number of properties including whether the pipe is bidirectional, whether it comprises a number of lower-speed pipes and its required and actual states.

SNMP Service

One area of the Network Model of particular interest for interoperability is the SNMP sub-tree. Given the ubiquity of SNMP management systems, it is likely that WBEM will often have to co-exist on the same device as an SNMP agent. In this case it could be useful for the WBEM operator to be able to manage the local SNMP agent.

I have illustrated some of the relevant classes in Figure 6.6. Note that, in order to show clearly how classes are associated with CIM_SNMPService, I have used a nonstandard manner of drawing the association: as a dotted line to indicate inheritance from the solid lines.

The heart of the SNMP interworking is the CIM_SNMPService class. Even though this class has neither properties nor members, it ties together the other classes associated with a particular SNMP implementation.

In order to access an SNMP agent you need to know a password, known in SNMP-speak as a Community String. This is held, together with details of the authority that an operator using the Community String has (i.e., read-only access, read-and-write access), in one or more instances of the CIM_SNMPCommunityString class. Instances of this class are tied to the CIM_SNMPService through the CIM_ServiceAccessBySAP association.

An alarm or indication is known, by SNMP experts, as a Trap. Instances of the CIM_SNMPTrapTarget class allow you to specify remote addresses to which SNMP traps should be sent. Again, this class is associated with an CIM_SNMPService instance, this time through the CIM_ServiceSAPDependency association. The CIM_SNMPTrapTarget class has a number of properties, including another Community String, mainly concerned with locating the remote server: a host address and the UDP port to which the traps are to be delivered (defaulting to port 162).

These classes tie the destination for the SNMP traps to the SNMP service. It is also necessary to define the origin of the traps: what is generating them? The originator is modelled as a CIM_ProtocolEndpoint, thus providing its IP address.

I give more information about possible SNMP to WBEM/CIM migration routes in Chapter 14.

QoS Service

There is one further significant CIM_Service defined in the Network Model: Quality of Service (QoS). This naming again reflects the origin of the Network Model in the packet- rather than circuit-switching area. In the packet world, Quality of Service is not a characteristic of any and all services, it is a shorthand for a particular service: the service that allows different customers to have their packets treated in different ways as they pass through IP routers. If we were both customers of a particular network operator, by paying more, you could ensure that your packets took precedence over mine in various queues in the routers and you would then have bought a "Quality of Service" Service.

In preparing the QoS submodel, the DMTF has worked closely with the IETF and at the time of writing, an IETF standard is due to emerge containing details of a model common to the DMTF and the IETF. This is currently in draft form as "Information Model for Describing Network Device QoS Datapath Mechanisms."

The Quality of Service submodel divides the components of QoS into two: the basic mechanisms for marking, dropping and metering IP packets which are collectively known as Conditioning Services and

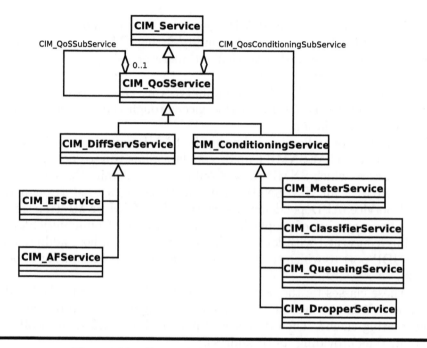

Figure 6.7   Part of the QoS Service Hierarchy

the aggregation of these into a service sold to a particular customer, for example, through Differentiated Services (RFC2474).

The two classes CIM_DiffServService and CIM_ConditioningService are shown in Figure 6.7 with a few of their subclasses. CIM_DiffServService, in particular, is subclassed into CIM_AFService and CIM_EFService to describe Assured Forwarding (AF) and Expedited Forwarding (EF) respectively. Because Expedited Forwarding does not actually require any more modelling than can be supplied by the CIM_ClassifierService and CIM_DiffServService classes, its use has now been deprecated; I have included it here because anyone knowledgeable in this area hearing of a class for the AF service would be waiting for the other shoe to fall, "What about EF service?"

We would expect that a service such as DiffServ would make use of the component services (packet metering, classifying, and dropping) and, as expected, such an aggregation does exist, CIM_QoSConditioningSubService, which aggregates the lower-level components into the higher-level service.

Equally, different DiffServ components might be combined to form a composite service for a particular customer. Thus the service "Customer X gets bronze service" would map through the CIM_QoSSubService aggregation to lower-level services which could be based on the various conditioning services reached through CIM_QosConditioning-SubService.

Apart from considering the services, there is one other useful classification within the Network Model: by administrative domain. The term "administrative domain" is used as a general term to mean some collection of network elements which together form a coherent whole and which can be combined with other administrative domains to create a higher-level domain. In the simplest case an administrative domain may be geographical: the Ontario domain and the Québec domain combining to form the East Central domain. The term is, however, more abstract than this and could refer to any hierarchical grouping.

Summary of the Network Model

The Network Model provides the infrastructure for managing network services: primarily routing, forwarding, SNMP, and QoS.

The Policy Model

The most important question here, of course, is, "What is meant by the term 'policy'?" A political party may have a policy of fiscal conservatism. I may have a policy of never going near small children, particularly grubby ones. Are these policies which can be modelled using the Policy Common Model? The answer is, only if they can be expressed in the form

```
IF (cond1 AND cond2) OR (cond1 and cond3) THEN action
```

It might be hard to express fiscal conservatism in this manner, but in principle, my aversion to children could be expressed as:

```
IF ((age lessThan 10) AND (person is grubby)) THEN avoid
```

Of course, I may wish to make exceptions to a rule: perhaps I want to avoid all grubby children except my own, George. I could then add a rule which says that:

```
IF (child is George) THEN provide bath
```

If George is under 10 years old and grubby, then both rules would be true and some mechanism must be defined to choose between them: do I avoid him or provide a bath? The answer, of course, is that I must be able to prioritise the rules and this is something that CIM provides.

More usefully in a telecommunications network one can imagine policies of the type

```
IF (customerClass is 1) AND
          (currentTrafficLevel lessThan 45 percent)
          THEN accept connection
```

Rules such as these would typically define the Service-Level Agreement (SLA) between a supplier (carrier) and a customer: "I will carry your traffic under these conditions."

The DMTF is working with the IETF to produce a formal description of a policy. This collaboration has produced RFC3460, which is entitled "Policy Core Information Model (PCIM) Extensions," and this, read in conjunction with the *mof* code, probably provides the best description of policies. RFC3460 itself is an update of RFC3060, entitled "Policy Core Information Model," and, as an introduction, RFC3060 is probably the better specification to read: it contains more description of policies without details of their application to particular IP routing elements.

Another useful document to help you wade through the policy swamp is RFC3198 entitled "Terminology for Policy-Based Management."

Figure 6.8 shows a small part of the Policy Common Model. Remember that the open diamonds represent aggregation (refer back to section 5 if necessary): a CIM_PolicyRule aggregates CIM_PolicyConditions (IF X IS Y ...) and CIM_PolicyActions (...THEN Z).

Although CIM_PolicyRule has many properties, I have chosen to include only two in Figure 6.8: ConditionListType and Enabled. ConditionListType is an enumerated variable which can take two values: disjunctive normal form (DNF) or conjunctive normal form (CNF). These are impressive terms to describe how the conditions are to be combined: in either case the conditions are grouped (by using the GroupNumber property on the CIM_PolicyConditionInPolicyRule association— see Figure 6.8). Assume that a rule has 6 conditions with the GroupNumber and ConditionNegated properties as shown in Table 6.1.

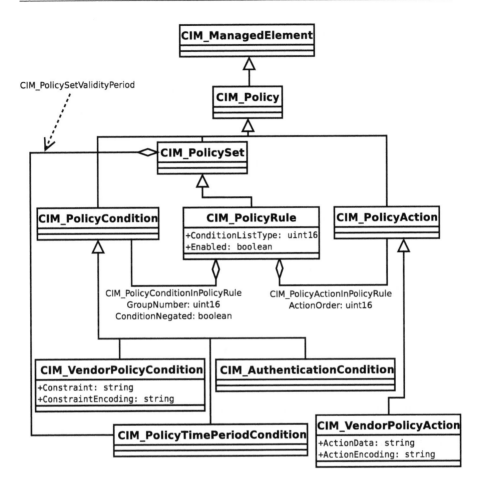

Figure 6.8   Part of the Policy Hierarchy

Condition	GroupNumber	ConditionNegated	Example
C1	45	False	Tuesday is the day before Wednesday
C2	43	False	$2 + 5 = 6$
C3	76	True	$3 \times 5 = 16$
C4	45	False	Canada lies to the north of the USA
C5	45	True	The capital of Canada is Toronto
C6	43	False	France is a monarchy

Table 6.1   Conditions

If the rule had the property of being expressed in Disjunctive Normal Form (DNF), then the conditions would be interpreted as

$$\overbrace{(C_1 \text{ AND } C_4 \text{ AND NOT } C_5)}^{\text{Group 45}} \text{ OR } \overbrace{(C_2 \text{ AND } C_6)}^{\text{Group 43}} \text{ OR } \overbrace{(\text{NOT } C_3)}^{\text{Group 76}}$$

whereas if the rule were in Conjunctive Normal Form (CNF) then the conditions would be interpreted as

$$\overbrace{(C_1 \text{ OR } C_4 \text{ OR NOT } C_5)}^{\text{Group 45}} \text{ AND } \overbrace{(C_2 \text{ OR } C_6)}^{\text{Group 43}} \text{ AND } \overbrace{(\text{NOT } C_3)}^{\text{Group 76}}$$

In each case, the NOT term arises from the ConditionNegated values. Given the example statements in the right-hand column of Table 6.1,[§] the Disjunctive Normal Form version would evaluate to

```
      (T and T and (not F)) or (F and F) or (not F)
i.e.  (T and T and T)       or (F and F) or (T)
i.e.  (T)                   or (F)       or (T)
```

which evaluates to TRUE (because both the first and third terms are TRUE). The Conjunctive Normal Form version, on the other hand, would evaluate to

```
      (T or T or (not F)) and (F or F) and (not F)
i.e.  (T or T or T)       and (F or F) and (T)
i.e.  (T)                 and (F)      and (T)
```

which evaluates to FALSE (because the second term is False).

This example illustrates the flexibility you have when combining conditions in a CIM_PolicyRule: you can negate any particular condition and then combine the resulting conditions in a flurry of ORs and ANDs.

The remaining question is how to express the conditions and actions. What language is used? We will start with conditions. As you can see in Figure 6.8, the common model defines three subclasses of CIM_PolicyCondition:

[§] Toronto, by the way, is *not* the capital of Canada—Ottawa is.

- CIM_PolicyTimePeriodCondition which allows you to specify time periods. For example, you could specify a condition that is true from 1st January 2005 to 17th June 2006, but only between 08:00 and 23:00 and only on Thursdays. Like any other condition, this can be combined (after being negated if necessary) with other conditions using OR and AND.

- CIM_AuthenticationCondition is designed to hold a condition defining how an operator is to be authenticated. You can consider it as a condition of the type "If the operator has been validated in accordance with X ..." To this end it has properties defining the type of validation (Shared Secret, Biometric {using mark, finger, voice, retina, DNA, EEG}, Public/Private Key Pair, Kerberos Ticket, Physical Credential, etc.) and the parameters and addresses required to complete the authentication. I discuss the use of this type of condition when considering the User Model below.

- CIM_VendorPolicyCondition which allows you to specify any condition, in a language of your choice. The class contains two fields which you can use to define the condition: Constraint and ConstraintEncoding. The DMTF has not specified a particular language and so any standard constraint language could be used. This is kept general by putting the OID of the language (e.g., 1.2.100.200) into the ConstraintEncoding field. Of course, your management software will have to include an interpreter to evaluate the condition in whatever language you have chosen and determine whether it is true or false.

    If you are using models from another source, your software could query the ConstraintEncoding property to decide whether it can handle the language or not. There may even be a policy to be followed if the software cannot handle the condition language (although encoding that policy might be hard!).

Because most conditions are likely to be simple ("property X has a value of Y"), RFC3460 subclasses CIM_PolicyCondition with an additional class, SimplePolicyCondition, which is designed to express precisely this type of condition:

```
DestinationPort MATCH {'80', '8080'}
DestinationPort MATCH {'1 to 255'}
DestinationPort MATCH '80'
```

Having grasped the encoding of conditions, you now implicitly understand the encoding of actions—it uses the same technique in that you get to choose the language and then specify an action and language in the ActionData and ActionEncoding properties of CIM_Vendor-PolicyAction. Because most actions are also likely to be simple ("set property X to the value Y"), RFC3460 subclasses CIM_PolicyAction with a new class, SimplePolicyAction, which is designed to express precisely this type of action (does this sound familiar?):

`Set Status to 'ON'`

As I have described them so far, there are two fundamental problems associated with policies:

- There could be an enormous number of them in a system. Maintaining these would be tedious and so the DMTF and IETF have introduced the concept of a *role* which I discuss below.
- There could be inconsistencies between the conditions or the conditions, when combined, could be inconsistent with the actual system resources ("Joe can have 50 percent of the bandwidth and Bill can have 70 percent"). Detecting such inconsistencies is a research problem and Cisco Systems is sponsoring work in this area at Imperial College, London. A recent paper, "Using CIM to Realise Policy Validation within the Ponder Framework," describes the application of the research to the DiffServ policies in CIM.

To address the first of these problems, the potentially enormous number of policies, each managed element can be assigned one or more roles. Policies can then be associated with roles rather than individual elements. RFC3460 defines a role as follows:

> A role is a type of attribute that is used to select one or more policies for a set of entities and/or components from among a much larger set of available policies.

Shakespeare said that "one man in his time plays many parts." Had he had the advantage of having read this book and RFC3460, he would have said that "one man in his time plays many roles." Sometimes he plays the role of father, at other times the role of taxi driver and at other times the role of barbecue chef. There might be policies appropriate to all fathers, another to all barbecue chefs.

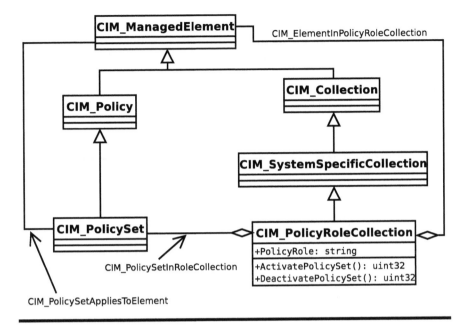

Figure 6.9   Roles in the Policy Hierarchy

A more useful example might relate to a company which has an IP network connecting its sites. In larger sites there might be separate telephony equipment (PBXs¶) and routers, but in smaller sites there might be devices combining the roles of PBXs and IP routers. This company might want to specify a policy relevant to all devices playing the part of an IP router, including both the dedicated routers and the PBXs acting as routers. There might be another policy relevant only to voice switching equipment. Another policy might be associated with both voice switching equipment *and* IP routers.

Were it not possible to associate a policy with a role, then it would have to be associated individually with each type of device, a much more time-consuming task.

I show part of the DMTF model dealing with Policy Roles in Figure 6.9. The class CIM_PolicyRoleCollection is an aggregation of CIM_PolicySets, each of which, as shown in Figure 6.8, *is a* CIM_Policy. CIM_PolicyRoleCollection has a property PolicyRole which is a string labelling the role (for example, "IP Routing") and must be associated with exactly one CIM_System.

CIM_PolicyRoleCollection is a sort of clearing-house between CIM_ManagedElements and CIM_PolicySets: it says that a particu-

---

¶ Private Branch eXchange: A small telephone exchange typically used within a building.

lar managed element *may* be subject to certain policies. Note the word "may." This is the power of roles—they can be defined but not activated until triggered. To activate a particular policy, CIM_PolicyRole-Collection provides a method ActivatePolicySet() which, when invoked, creates an association, CIM_PolicySetAppliesToElement, directly between the Managed Element concerned and the Policy Set. A similar method, DeactivatePolicySet(), removes this association.

Summary of the Policy Model

A policy is simply a set of rules, each consisting of a number of conditions and associated actions.

The User Model

Like the Network Model, the User Model is very large and somewhat incomplete. Between version 2.7 and 2.8 of the CIM schema it has been significantly simplified; unfortunately the white paper (DSP0139) which describes it, has not (at the time of writing) been updated to reflect these changes. If you read DSP0139, then ignore references to the class CIM_UsersAccess (no apostrophe) which used to be pivotal and is now deprecated.

The User Model is about managing the users of a device: their names, addresses, credentials and authorities. Users in this sense are either humans or programs using the services provided by the device to be managed. The users of a PBX include those people who use telephones, those people who access their voice mail, those computers which connect to it to extract billing information, etc.

We can therefore expect to find the following concepts somewhere in the User Model:

- The concept of a user with a name, address, etc.
- The concept of an account which the user may have on a particular system; we would expect the account to be "weak to the system" in the sense described on page 68.
- The concept of some form of authentication to ensure that the user is who he says he is (anything from the knowledge of a password to a full DNA analysis).
- The concept of some form of authorisation saying what a particular user, once authenticated, may do.

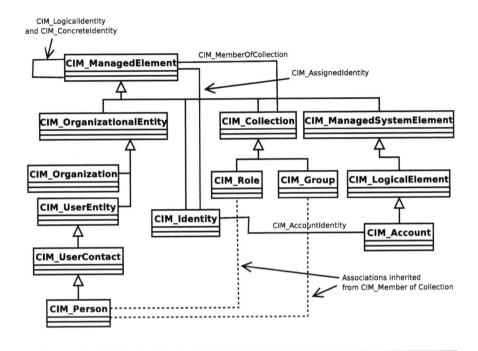

CIM_LogicalIdentity
and CIM_ConcreteIdentity

CIM_MemberOfCollection

CIM_AssignedIdentity

CIM_ManagedElement

CIM_OrganizationalEntity   CIM_Collection   CIM_ManagedSystemElement

CIM_Organization   CIM_Role   CIM_Group   CIM_LogicalElement

CIM_UserEntity

CIM_Identity   CIM_AccountIdentity   CIM_Account

CIM_UserContact

Associations inherited
from CIM_Member of Collection

CIM_Person

Figure 6.10   Part of the User Model

As we review the User Model at high level in the remainder of this section, bear in mind that, as always in CIM, the model is about the management of things, not about their implementation. This is particularly true in the case of users—the CIM system is not concerned about authenticating users and checking that they do not exceed their authority; it is concerned with managing the information that allows that authentication to take place. There is, of course, a related issue of ensuring that operators accessing the WBEM server itself are correctly authenticated and not allowed to do more than their authority permits. I describe this aspect of user management starting on page 142.

The Concept of a User

Of course, managing users is not new. Standards such as the ITU-T's X.500 have existed for a decade or more and contain lists of properties of users: title, surname, common name, street address, province or state, locality, country, organisation, etc. X.500 is considered to be too heavy for use in an enterprise and many of its features have been extracted into the Lightweight Directory Access Protocol (LDAP). LDAP is defined in a number of IETF RFCs; the complete list being contained in RFC3377. Note that, in spite of its name, LDAP does not just describe the protocol for accessing and updating user information; it also includes details of the properties that are typically exchanged.

When the designers of CIM's User Model came to define the class structure and properties of users they naturally made use of the work already done in X.500 and LDAP. This ensures compatibility with LDAP.

The details of the mapping of LDAP to CIM is described in the DMTF white paper number DSP0123, "LDAP Mapping Specification." Take, for example, the property "businessCategory." This is taken from standard X.520 and is defined in RFC2256 as follows:

```
businessCategory
This attribute describes the kind of business performed
by an organization.
( 2.5.4.15 NAME 'businessCategory'
  EQUALITY caseIgnoreMatch
  SUBSTR caseIgnoreSubstringsMatch
  SYNTAX 1.3.6.1.4.1.1466.115.121.1.15{128} )
```

This defines a property called businessCategory which is a string (believe it or not, "1.3.6.1.4.1.1466.115.121.1.15" means "string") of up to 128 characters. This maps into a property of the CIM_Person class:

```
[MaxLen (128), Description (
   "This property describes the kind of business "
   "performed by an organization.") ]
string BusinessCategory;
```

where the concept of a string is captured somewhat more succinctly by using the word "string."

There is one more important classification of users handled by the User Model. Users, particularly human ones, are gregarious and gather in groups: they are members of a department along with other people; they have a certain role in common with other people (e.g. System Administrator). Such groups can naturally be modelled within CIM as CIM_Collections (as described on page 95) and the membership relationship by the CIM_MemberOfCollection association.

A User's Identities and Accounts

There are two different views of user management:

1. The user looks at the system and asks, "Which parts of the system can I access?"
2. The system looks back at the user and says, "Who is this person, where is she, what are her credentials, can she prove that she's who she says she is?"

The User Model covers both of these viewpoints—the former starting with an instance of CIM_UserEntity, the latter with CIM_Account and its close friend CIM_Identity.

CIM_Identity is really the centre of the authentication world—it says that the owner of this identity has (or has not) had its credentials checked. Once the credentials are checked, it provides a level of trust and degree of access for the identified object.

As you can see from Figure 6.10, an identity, whether checked or not, can be associated with any CIM_ManagedElement through the CIM_AssignedIdentity association.

There is another class, CIM_Account, which also has aspects of verifying an identity, particularly a person's UNIX-style identity. In some respects, CIM_Identity and CIM_Account overlap and, had history been different, they would probably not exist as separate classes. As it is, they are closely associated by the CIM_AccountIdentity association.

Any user, and in particular a human user, a CIM_Person, can now have one or more associated identities (through instances of CIM_AssignedIdentity) and thereby an instance of CIM_Account (through CIM_AccountIdentity).

CIM_Account is a useful class to reach because it has associations, as illustrated in Figure 6.11, which connect it to a particular system. This completes the chain: the user has accounts of various types on various systems.

Authentication

The description so far has avoided the two crucial issues of authentication and authorisation: "Is this user who he says he is?" and "Is this user allowed to do this operation?"

Authentication can be thought of as a matter of policy: IF this user knows this password (or has that iris pattern in his right eye), THEN consider him authenticated. I have already mentioned (page 111) the CIM_AuthenticationCondition subclass of CIM_PolicyCondition. It is by satisfying the condition specified therein that a user becomes officially authenticated and gets the CurrentlyAuthenticated flag set in its instance of CIM_Identity.

Although performing the actual credential check is outside the scope of the WBEM management system, an authentication service

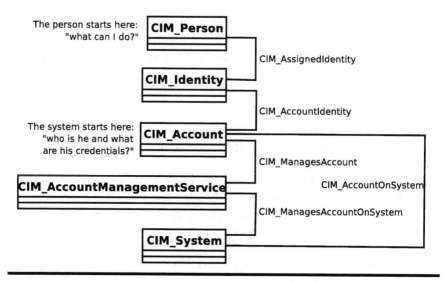

Figure 6.11   Users: The Chain of Associations

(CIM_AuthenticationService) is defined (see Figure 6.12) and a significant part of the User Model is dedicated to classes representing credentials—shared keys, public keys, biometrics, etc.

One point should, however, be clear: the Boolean property CurrentlyAuthenticated on an instance of CIM_Identity should generally not be writable by operators!

Authorisation

When a user is authenticated, he, she, or it is given certain privileges. Each privilege is described by an instance of the CIM_Privilege class.

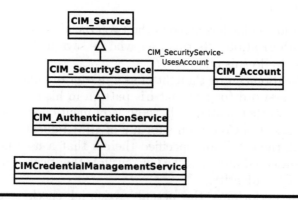

Figure 6.12   The AuthenticationService

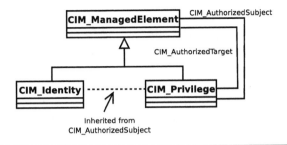

Figure 6.13   CIM_Privilege

This contains information about the type of privilege (write access, read access, execute access) it affords to the user.

An instance of CIM_Privilege sits between a user's identity and a resource to which he has privileges—see Figure 6.13. CIM_Authorized-Subject is the link to the identity and CIM_AuthorizedTarget the link to the resource.

Summary of the User Model

The User Model is designed to provide a set of classes, compatible where possible with LDAP, within which details of system users, their authentication mechanisms, and authority levels are stored.

Frequently Asked Questions

FAQ 22   *The common models are large and difficult to understand. Can I simply ignore them and write my own models?*

The straight answer is "yes." Nothing breaks in the WBEM architecture if the DMTF models are not used; the classes contained in these models are simply loaded into the WBEM server by a client (the *mof* compiler) just like any other classes.

However, this is something that you might like to think about *very* carefully: the common and core models *are* complex, but not (completely) out of awkwardness on the part of the writers—they are complex largely because the areas that they cover are complex. If you have a system for which there is, as yet, no common model then you might still like to consider taking the core model from the DMTF. Using this will at least give your model basic structure and possibly even allow you to make your model standard in the future. In addition, most devices are likely to be able to make use of other parts of the common models: do you have hardware, software, users, operating systems which must be managed? If so, then the common models might be useful to you.

There is one further reason for using the standard models. Recently, I was creating an instance of CIM_EthernetPort and assumed that there would be a property for the MAC address of the port. When I looked at the common model, I was puzzled to see that an array was defined to contain the MAC addresses. I am no expert on Ethernet ports and spoke with a colleague who patiently explained to me the conditions under which several MAC addresses would be associated with an Ethernet port. Had I created the model of an Ethernet port myself, I would, out of ignorance, have defined a single MAC address. Sometime in the future this could have given me trouble. By using the common models you are also using the expertise of the people who wrote them and, in many areas, that expertise might be deeper than yours.

FAQ 23 *How does describing a user and his account in the User Model actually cause the validation of the user to occur?*

It does not. CIM is a means of describing management data in a standard and, one hopes, logical form. The User Model allows you to store the information associated with the user's passwords, access lists, etc. in a standard form. It does not provide the code actually to authenticate the user and control his or her access.

INTERFACES

INTERFACES

Chapter 7

The Client/Server Interface

Introduction

Chapters 11 and 13 describe WBEM servers, listeners, and clients in detail. An element common to these entities is the interfaces between them. This and the next chapter describe those interfaces, setting the scene for the later chapters.

It is the intent of the WBEM architecture that the WBEM clients and listeners have no knowledge of the providers and the manner in which the information that they are requesting and using is obtained. The protocol is designed to isolate the two parts, with the WBEM server acting as a broker between the two. For example, the operating system class has a property, `LastBootUpTime`, which reasonably holds the time that it was last booted. The manner in which this information is actually extracted from a running operating system by a provider differs, depending on the actual operating system: Linux, HPUX, AIX, MacOS, Z/OS, or Microsoft Windows. The WBEM client should be completely unaware of this, simply requesting the value of the `LastBootTime` property of the appropriate instance of the CIM_OperatingSystem class.

The WBEM server lies at one end of this interface but, as Figure 7.1 shows, two types of entity lie at the other end:

1. A number of WBEM clients—the code which encapsulates the operators' requests and commands and passes them to the WBEM server for action.

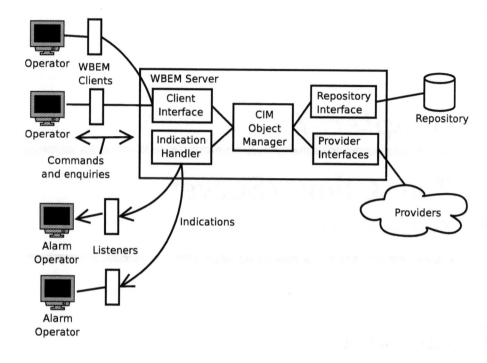

Figure 7.1   The WBEM Client/WBEM Server Interfaces

2. A number of WBEM listeners—the code which receives information about events (indications) that have occurred in the managed system.

The major distinction between these two is the stimulus which starts the interaction—in the former case it is the operator or higher-level management system entering a command to which the WBEM server responds and in the latter case it is an event occurring in the system. There is, however, one slightly more subtle difference between these two interfaces—in the case of the client/server interface it can be assumed that both ends understand CIM and have access to the CIM model; in the case of the server/listener interface it cannot be assumed that the listener has any understanding of CIM. Typically a listener will be handling events from different management systems and so any information sent to it must be carefully de-CIMed.

As part of its WBEM initiative, the DMTF has standardised these interfaces in a set of documents:

- **CIM Operations over HTTP, DSP0200** which defines both the client/server and the server/listener interfaces. It defines the operations clients can ask a WBEM server to perform and the format of messages exported by the WBEM server to listeners.
- **Representation of CIM in XML, DSP0201** which defines the precise XML syntax of the messages a client would use to invoke a function on a WBEM server. The document defines an XML grammar, written in DTD.*
- **XML Document Type Definition** which, for convenience, contains the DTD grammar extracted from document DSP0201.

All of these documents are freely available from the DMTF Web site.

Although the DMTF has standardised the interface between the WBEM client and the WBEM server using XML, note that most WBEM server implementations additionally provide a higher-performance binary interface: essentially the same functions but without the XML encoding. This is a typical trade-off—use the XML interface for compatibility, use the binary interface for higher performance.

The remainder of this chapter deals with the details of the client/server interface; Chapter 8 deals with the listener interface.

A Survey of the Client/Server Interface

Figure 7.2 shows the various layers through which an operator's request passes and through which the response returns when a managed object is accessed.

A command from the operator or higher-level management system originates with the application logic which handles graphical screens, interpretation of command-line interfaces, etc. The command is then mapped to the abstract object model. This is not a trivial activity and I discuss possible approaches in Chapter 13, in particular the section on Semantic Knowledge starting on page 248. Fundamentally, the Object Abstraction layer has to know that, when the operator enters a command to create a new OSPF service, for example, then this means the creation of instances of various CIM classes. This knowledge may be hard-coded, table-driven or driven from the model.

Once the actual CIM commands have been determined, they are encapsulated in CIM-XML and passed to the HTTP client. This is

---

* Document Type Definition.

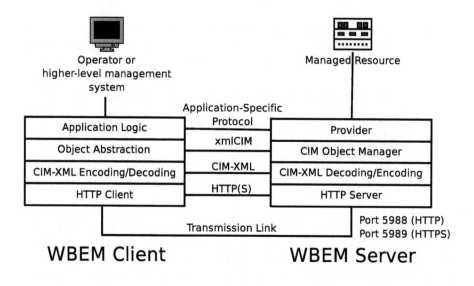

Figure 7.2   WBEM Client/WBEM Server Interaction

responsible for any negotiation required with the WBEM server and the correct transfer of the request.

Once the request reaches the WBEM server, it is passed from the HTTP server to have the CIM-XML reconstituted into CIM commands. The CIM Object Manager (CIMOM) examines these and determines how they should be handled: by the CIMOM itself, with reference to the schema stored in the repository, or by a provider. If it is passed to the provider, then the provider accesses the real device or service to retrieve the requested information or carry out the requested configuration.

The response from the provider follows the reverse path back to the operator.

The Connection/Disconnection Phase

Establishing the connection between the WBEM client and WBEM server uses the standard interchanges associated with HTTP or HTTPS. These comprise the establishment of a TCP session and necessary authentication (see page 142).

CIM Message Transfer

Assuming that the WBEM client has connected to the WBEM server, I now address three questions about the subsequent exchange:

1. What requests can be made by a WBEM client to a WBEM server?
2. How are those requests encoded?
3. How are the encoded requests actually transported to and from the WBEM server?

The protocols used to exchange information between a WBEM client and WBEM server are illustrated in Figure 7.2. The WBEM client prepares requests and commands in xmlCIM and logically exchanges these with the WBEM server. In practice, in order to encode the message, the WBEM client passes the xmlCIM to a CIM-XML client which itself inserts the request or command into an HTTP message, which is then transferred to an HTTP server, typically on the device being managed. The DMTF has reserved HTTP ports 5988 and 5989 for HTTP and HTTPS, respectively. Once received from the HTTP transport, the request is unpacked and passed as an xmlCIM packet to the CIM Object Manager (CIMOM).

The request, command, or response actually transmitted from the client to the server is formally known as a CIM Operation Message: it is effectively a message exchanged between two CIM-aware entities. Figure 7.3 illustrates a simple CIM-XML operational message: one which carries a query from a WBEM client to a WBEM server (this example, although generic, was taken from the openPegasus test suite and is therefore subject to the licence described on page 313. If you are XML-literate, you will see that this message contains a request for the WBEM server to enumerate (i.e., "list") the names of instances of the class Linux_CDROMDrive in the "root/cimv2" namespace. In simpler terms, this is a request from an operator to the WBEM server for a list of all CDROM drives being supported by a particular Linux operating system.

In order to define the contents of the CIM messages across the client/server interface, it is best to start with the list of operations that a client might wish to perform. These are divided into two basic types: intrinsic and extrinsic. The XML actually encodes the type of operation, using IMETHODCALL for an intrinsic operation as in Figure 7.3 and METHODCALL for an extrinsic one. I describe the intrinsic operations in the next section and extrinsic ones starting on page 140.

```
<?xml version="1.0" encoding="utf-8"?>
<CIM CIMVERSION="2.0" DTDVERSION="2.0">
  <MESSAGE ID="34422" PROTOCOLVERSION="1.0">
    <SIMPLEREQ>
      <IMETHODCALL NAME="EnumerateInstanceNames">
        <LOCALNAMESPACEPATH>
          <NAMESPACE NAME="root"/>
          <NAMESPACE NAME="cimv2"/>
        </LOCALNAMESPACEPATH>
        <IPARAMVALUE NAME="ClassName">
          <CLASSNAME NAME="Linux_CDROMDrive"/>
        </IPARAMVALUE>
      </IMETHODCALL>
    </SIMPLEREQ>
  </MESSAGE>
</CIM>
```

Figure 7.3   A Simple CIM-XML Message

Intrinsic Methods

Intrinsic methods are those designed to be built into the WBEM server (but may not all be supported by every WBEM server—see the description of the Interop common model on page 250 for the method a WBEM client may use to discover precisely which operations a WBEM server supports). Intrinsic methods are oriented towards the manipulation of the model itself and include methods to retrieve, delete, create, list ("enumerate" in CIM-speak), and generally manipulate classes, instances, associations, and qualifiers.

A full description of each of the intrinsic methods is given in the DMTF's document DSP0200 and I do not intend to repeat it here. Instead, I give a short list of the intrinsic methods in Table 7.1 and details of a few representative operations below (indicated by **bold** type in the table). I list these methods again in Figure 13.2 on page 253 where they are grouped into sets which different WBEM servers may or may not be able to handle.

Notice that although getProperty and setProperty appear in Table 7.1, they are effectively redundant since getInstance and modifyInstance both allow a subset of properties (including one) to be specified.

Class-Oriented Methods

GetClass	Return the specification of a particular class
CreateClass	Create a new class
DeleteClass	Remove a class
ModifyClass	Change the specification of a class
EnumerateClasses	List all classes with particular characteristics
EnumerateClassNames	List the names of all classes with particular characteristics

Instance-Oriented Methods

GetInstance	Return an instance defined by a unique combination of keys
CreateInstance	Create an instance of a particular class
DeleteInstance	Remove an instance
ModifyInstance	Modify one or more properties of an instance
EnumerateInstances	List some or all of the properties of instances with particular characteristics
EnumerateInstanceNames	List the names of instances with particular characteristics
GetProperty	Get the value of one property in an instance
SetProperty	Set the value of one property in an instance

Qualifier-Oriented Methods

GetQualifier	Return a qualifier from a given namespace
SetQualifier	Create or modify a particular qualifier in a given namespace
DeleteQualifier	Remove a qualifier from a namespace
EnumerateQualifiers	List qualifiers in a namespace

Association-Oriented Methods

Associators	Return a list of instances associated with a particular instance or classes associated with a particular class
AssociatorNames	List the names of instances associated with a particular instance or classes associated with a particular class
References	Return a list of associations for a particular class or instance
ReferenceNames	Return a list of names of associations for a particular class or instance

Miscellaneous Methods

ExecQuery	Execute a (potentially complex) database-style query on the classes and instances

Table 7.1 The Intrinsic Methods

Enumerating (i.e., listing) Instances

Because clients often need access to particular properties of instances, I will describe this operation very fully. This will then give a flavour for the other operations which I will describe more superficially.

The function EmumerateInstances is defined as follows to allow a client to retrieve a list of all instances satisfying certain criteria:

```
<namedInstance>*EnumerateInstances (
        [IN] <className> ClassName,
        [IN,OPTIONAL] boolean LocalOnly = true,
        [IN,OPTIONAL] boolean DeepInheritance = true,
        [IN,OPTIONAL] boolean IncludeQualifiers = false,
        [IN,OPTIONAL] boolean IncludeClassOrigin = false,
        [IN,OPTIONAL,NULL] string PropertyList [] = NULL)
```

I have used here the pseudocode employed by the DMTF in their specifications. It is reasonably self-explanatory with the possible exception of IN, OUT, and OPTIONAL. These define whether the parameter is passed into the WBEM server (IN) or whether the WBEM server returns a value in the parameter (OUT). OPTIONAL means that the parameter does not have to be present—the default value given after the equals sign being taken if it is absent.

This call effectively says, "Give me a list of properties X, Y, and Z for all instances of class C." More formally, you need to provide:

ClassName to define the class for which instances are required. Note that this is the full class name, including namespace.

LocalOnly and DeepInheritance to define precisely the properties that are to be returned. Table 7.2 shows that the properties from the specified class are always candidates for return but that properties from subclasses and superclasses of the specified class can be included or excluded as required.

As the use of these two parameters might not be intuitive (!), consider Figure 7.4.

Assume that we issue an EnumerateInstances call for the class B and assume that we set PropertyList to accept everything (i.e., leave it as NULL). Then the actual properties returned are as illustrated in Table 7.3.

IncludeQualifiers to specify whether or not qualifiers for each instance and property are to be returned. If support for qualifiers on instances is dropped by the DMTF then this parameter may also disappear.

| LocalOnly | DeepInheritance | Class | .. Properties in .. | |
			Subclasses	Superclasses
True	True	Yes	Yes	No
True	False	Yes	No	No
False	True	Yes	Yes	Yes
False	False	Yes	No	Yes

Table 7.2 **Properties Retrieved by** `EnumerateInstances()`

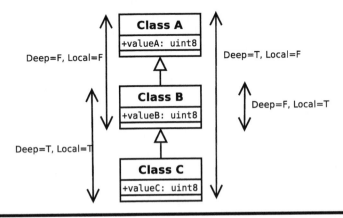

Figure 7.4   **The DeepInheritance and LocalOnly Parameters**

`IncludeClassOrigin` to specify whether the name of the class in which the property or method was defined (possibly a superclass of this one) should be included in the response.

`PropertyList` to specify which properties of the instances should be returned. If you leave this parameter as NULL then all properties are returned. If the list contains names which are not properties of the class then the WBEM server is required to ignore those names and process the other items in the list.

LocalOnly	DeepInheritance	Properties Retrieved
True	True	ValueB and ValueC
True	False	ValueB
False	True	ValueA, ValueB and ValueC
False	False	ValueA and ValueB

Table 7.3   **Behaviour of LocalOnly and DeepInheritance**

Note that PropertyList, LocalOnly, and DeepInheritance all define the properties that are to be returned. The properties actually returned are the those common to all three parameters.

If successful, **EnumerateInstances** returns the requested instances. Otherwise it returns one of the following status codes:

- CIM_ERR_ACCESS_DENIED to indicate that the WBEM client was not authorised to access the WBEM server with this request. Deciding whether the operator does or does not have the authorisation to carry out an operation lies primarily with the provider. This error might mean something as simple as the operator not having the right to access a particular file needed for the operation.
- CIM_ERR_NOT_SUPPORTED to indicate that the WBEM server does not support this function. This should not occur if the WBEM server has previously supported the function but it may occur the first time that an intrinsic method is invoked. A WBEM client can, to some extent, ask the WBEM server what methods it supports but the standard allows the WBEM server to lie—see page 250.
- CIM_ERR_INVALID_NAMESPACE to indicate that the specified namespace does not exist.
- CIM_ERR_INVALID_PARAMETER to indicate that at least one parameter is missing, duplicated, unrecognised or is otherwise incorrect.
- CIM_ERR_INVALID_CLASS to indicate that the specified class does not exist.
- CIM_ERR_FAILED to indicate that some other, unspecified error occurred.

If multiple errors occur then only the one earliest in this list is returned.

Sometimes it would be convenient for the provider handling an **EnumerateInstances** call from a WBEM client to return a different error code, indicating the type of error in an application-specific manner. At present, this is not possible, but a descriptive field is associated with each error and could be used by the provider to pass additional coded or free-text information.

The XML which would pass between the WBEM client and the WBEM server for an invocation of **EnumerateInstances** is given in Figure 7.5. Again, this is adapted from the openPegasus test suite and is therefore subject to the openPegasus licence. Notice again the IMETHODCALL parameter, this time with a name of EnumerateIn-

```
<?xml version="1.0" ?>
<CIM CIMVERSION="2.0" DTDVERSION="2.0">
 <MESSAGE ID="51007" PROTOCOLVERSION="1.0">
  <SIMPLEREQ>
   <IMETHODCALL NAME="EnumerateInstances">
    <LOCALNAMESPACEPATH>
     <NAMESPACE NAME="root"/>
     <NAMESPACE NAME="cimv2"/>
    </LOCALNAMESPACEPATH>
    <IPARAMVALUE NAME="ClassName">
       <CLASSNAME NAME="CIM_ComputerSystem"/>
    </IPARAMVALUE>
    <IPARAMVALUE NAME="DeepInheritance">
       <VALUE> FALSE </VALUE>
    </IPARAMVALUE>
    <IPARAMVALUE NAME="LocalOnly">
       <VALUE> FALSE </VALUE>
    </IPARAMVALUE>
    <IPARAMVALUE NAME="IncludeQualifiers">
       <VALUE> FALSE </VALUE>
    </IPARAMVALUE>
    <IPARAMVALUE NAME="IncludeClassOrigin">
       <VALUE> FALSE </VALUE>
    </IPARAMVALUE>
   </IMETHODCALL>
  </SIMPLEREQ>
 </MESSAGE>
</CIM>
```

Figure 7.5   Example XML for EnumerateInstances

stances. The parameters of the **EnumerateInstances** call follow in the remaining XML.

Enumerating (i.e., listing) Classes

The pseudocode for enumerating classes is as follows:

```
<class>*EnumerateClasses (
         [IN,OPTIONAL,NULL] <className> ClassName=NULL,
         [IN,OPTIONAL] boolean DeepInheritance = false,
         [IN,OPTIONAL] boolean LocalOnly = true,
```

```
[IN,OPTIONAL] boolean IncludeQualifiers = true,
[IN,OPTIONAL] boolean IncludeClassOrigin = false)
```

In this call:

- The `ClassName` parameter defines the class for the enumeration and, if it is missing, all classes within the namespace are chosen.
- `DeepInheritance` and `LocalOnly` are almost as complex as they are for `EnumerateInstances`:
  - If `DeepInheritance` is true, then all subclasses of the specified class are returned. If it is false, then only immediate child subclasses are returned.
  - If `LocalOnly` is true, then only CIM Elements (properties, methods, and qualifiers), defined or overridden within the definition of the selected class (i.e., not those of subclasses) are returned.
- `IncludeQualifiers` specifies whether or not the qualifiers for each selected class should be returned.
- `IncludeClassOrigin` specifies whether the CLASSORIGIN attribute should be returned. This field contains the name of the class in which the method or property was originally defined (possibly a superclass of the class being returned).

If successful, `EnumerateClasses` returns zero or more classes that meet the required criteria. If unsuccessful, `EnumerateClasses` returns one of the status codes described for `EnumerateInstances` above.

I give an example of the XML which might pass between the WBEM client and the WBEM server to invoke the `EnumerateClasses` method in Figure 7.6. Again note the use of the IMETHODCALL tag to indicate that this is an intrinsic method call and the SIMPLEREQ tag to indicate that only a single request is contained in the invocation. And, once again, this example is modified slightly from an openPegasus test program and so is subject to the openPegasus licence. If more than one method call were contained in the command then SIMPLEREQ would be replaced by MULTIREQ and any number of IMETHODCALLs and METHODCALLs could be included.

Creating an Instance

I am sure that the pattern for these WBEM client operations is becoming clear; a straightforward function defined by the DMTF in pseudocode, possibly with a couple of nonintuitive parameters such

```
<?xml version="1.0" encoding="utf-8"?>
<CIM CIMVERSION="2.0" DTDVERSION="2.0">
  <MESSAGE ID="42002" PROTOCOLVERSION="1.0">
    <SIMPLEREQ>
      <IMETHODCALL NAME="EnumerateClasses">
        <LOCALNAMESPACEPATH>
          <NAMESPACE NAME="test"/>
          <NAMESPACE NAME="cimv2"/>
        </LOCALNAMESPACEPATH>
        <IPARAMVALUE NAME="ClassName">
          <CLASSNAME NAME="CIM_ComputerSystem"/>
        </IPARAMVALUE>
        <IPARAMVALUE NAME="DeepInheritance">
          <VALUE> FALSE </VALUE>
        </IPARAMVALUE>
        <IPARAMVALUE NAME="LocalOnly">
          <VALUE> FALSE </VALUE>
        </IPARAMVALUE>
        <IPARAMVALUE NAME="IncludeQualifiers">
          <VALUE> FALSE </VALUE>
        </IPARAMVALUE>
        <IPARAMVALUE NAME="IncludeClassOrigin">
          <VALUE> FALSE </VALUE>
        </IPARAMVALUE>
      </IMETHODCALL>
    </SIMPLEREQ>
  </MESSAGE>
</CIM>
```

Figure 7.6   Example XML for EnumerateClasses

as `DeepInheritance` and `LocalOnly`, and then a very straightforward mapping into XML.

Following this pattern, creating an instance is defined in pseudocode as follows:

```
<instanceName>CreateInstance (
        [IN] <instance> NewInstance)
```

It is hard to think of a simpler interface, but, of course, the parameter itself is more complex this time. This can be most easily seen by looking at the XML for a call to CreateInstance; see Figure 7.7

(again taken from the openPegasus test suite). This example creates an instance of the CIM_IndicationFilter class. At present you do not need to understand what this class does—I explain it in detail on page 157—simply accept that such a class exists and has properties such as SystemName, SystemCreationClassName, and Query.

Again the XML shows that CreateInstance is an intrinsic call (IMETHODCALL) and the general structure of the XML is:

```
<IMETHODCALL NAME="CreateInstance">
 <LOCALNAMESPACEPATH>
  <NAMESPACE NAME="AAAAAA"/>
 </LOCALNAMESPACEPATH>
 <IPARAMVALUE NAME="NewInstance">
  <INSTANCE CLASSNAME="CIM_IndicationFilter">
   <PROPERTY NAME="XXXXXXXX" TYPE="YYYYYY">
          <VALUE>ZZZZZZZZ</VALUE>
   </PROPERTY>
  </INSTANCE>
 </IPARAMVALUE>
</IMETHODCALL>
```

where the PROPERTY tag is repeated for each property in the instance, each time giving the property's name, type, and value.

Deleting an Instance

With EnumerateInstances and CreateInstance well understood, you will find DeleteInstance trivial. Here is the pseudocode:

```
void DeleteInstance([IN] <instanceName> InstanceName)
```

and the associated XML example taken from the openPegasus test suite is in Figure 7.8. The XML this time simply contains those properties that are keys on a CIM_IndicationFilter so that the WBEM server can uniquely identify the instance to be deleted; see Frequently Asked Question 27 on page 148.

Traversing Associations

The Associators method and its companions References, ReferenceNames and AssociatorNames allow you to manipulate associations. Associations, however, are just normal classes and so can

```xml
<?xml version="1.0" encoding="utf-8"?>
<CIM CIMVERSION="2.0" DTDVERSION="2.0">
  <MESSAGE ID="53000" PROTOCOLVERSION="1.0">
    <SIMPLEREQ>
      <IMETHODCALL NAME="CreateInstance">
        <LOCALNAMESPACEPATH>
          <NAMESPACE NAME="root"/>
          <NAMESPACE NAME="cimv2"/>
        </LOCALNAMESPACEPATH>
        <IPARAMVALUE NAME="NewInstance">
          <INSTANCE CLASSNAME="CIM_IndicationFilter">
           <PROPERTY NAME="SystemCreationClassName" TYPE="string">
            <VALUE>CIM_UnitaryComputerSystem</VALUE>
           </PROPERTY>
           <PROPERTY NAME="SystemName" TYPE="string">
            <VALUE>server001.acne.com</VALUE>
           </PROPERTY>
           <PROPERTY NAME="CreationClassName" TYPE="string">
            <VALUE>CIM_IndicationFilter</VALUE>
           </PROPERTY>
           <PROPERTY NAME="Name" TYPE="string">
            <VALUE>ACNESubscription12345</VALUE>
           </PROPERTY>
           <PROPERTY NAME="SourceNamespace" TYPE="string">
            <VALUE>root/cimv2</VALUE>
           </PROPERTY>
           <PROPERTY NAME="Query" TYPE="string">
            <VALUE>
             SELECT Description, AlertType \
                FROM CIM_AlertIndication WHERE PerceivedSeverity = 3
            </VALUE>
           </PROPERTY>
           <PROPERTY NAME="QueryLanguage" TYPE="string">
            <VALUE>WQL</VALUE>
           </PROPERTY>
          </INSTANCE>
        </IPARAMVALUE>
      </IMETHODCALL>
    </SIMPLEREQ>
  </MESSAGE>
</CIM>
```

Figure 7.7   Example XML for CreateInstance

```xml
<?xml version="1.0" encoding="utf-8"?>
<CIM CIMVERSION="2.0" DTDVERSION="2.0">
  <MESSAGE ID="53000" PROTOCOLVERSION="1.0">
    <SIMPLEREQ>
      <IMETHODCALL NAME="DeleteInstance">
        <LOCALNAMESPACEPATH>
          <NAMESPACE NAME="root"/>
          <NAMESPACE NAME="cimv2"/>
        </LOCALNAMESPACEPATH>
        <IPARAMVALUE NAME="InstanceName">
          <INSTANCENAME CLASSNAME="CIM_IndicationFilter">
            <KEYBINDING NAME="SystemCreationClassName">
              <KEYVALUE VALUETYPE="string">
              CIM_UnitaryComputerSystem
              </KEYVALUE>
            </KEYBINDING>
            <KEYBINDING NAME="SystemName">
              <KEYVALUE VALUETYPE="string">
              server001.acne.com
              </KEYVALUE>
            </KEYBINDING>
            <KEYBINDING NAME="CreationClassName">
              <KEYVALUE VALUETYPE="string">
              CIM_IndicationFilter
              </KEYVALUE>
            </KEYBINDING>
            <KEYBINDING NAME="Name">
              <KEYVALUE VALUETYPE="string">
              ACNESubscription12345
              </KEYVALUE>
            </KEYBINDING>
          </INSTANCENAME>
        </IPARAMVALUE>
      </IMETHODCALL>
    </SIMPLEREQ>
  </MESSAGE>
</CIM>
```

Figure 7.8   Example XML for DeleteInstance

be manipulated with the class operations that we discussed above so these additional operations are convenient but not strictly required.

`Associators` and `AssociatorNames` are used to find out which instances are associated with a particular instance or which classes are associated with a particular class. The only difference between the operations is that `Associators` returns the whole object (using the strict definition of this word) and `AssociatorNames` returns only the names. The pseudocode for `Associators` is

```
<objectWithPath>*Associators (
    [IN] <objectName> ObjectName,
    [IN,OPTIONAL,NULL] <className> AssocClass = NULL,
    [IN,OPTIONAL,NULL] <className> ResultClass = NULL,
    [IN,OPTIONAL,NULL] string Role = NULL,
    [IN,OPTIONAL,NULL] string ResultRole = NULL,
    [IN,OPTIONAL] boolean IncludeQualifiers = false,
    [IN,OPTIONAL] boolean IncludeClassOrigin = false,
    [IN,OPTIONAL,NULL] string PropertyList [] = NULL )
```

This means, in effect, "please give me a list of all objects (instances or classes) which are associated with object Y." The additional properties allow the search to be pruned to include only objects which are associated through a specific association class or where the objects take particular roles in the association.

Note that `Associators`, `References`, `ReferenceNames`, and `AssociatorNames` differ from the other intrinsic methods in that they may operate on classes ("with what other classes is class A associated?") or on instances ("with what instances is instance B associated?"). Other intrinsic calls operate on either classes (e.g., `DeleteClass`) or on instances (e.g., `DeleteInstance`). If an `Associators`, `References`, `ReferenceName` or `AssociatorNames` request is made for a class then it is satisfied by the WBEM server; if the request is made for an instance then the WBEM server passes it to the appropriate provider if there is one.

`References` and `ReferenceNames` work similarly, returning not the objects at the far end of the references but the references themselves. These operations effectively answer the question, "By means of which associations is object A associated with anything?"

Executing Complex Queries

The `ExecQuery` method allows the WBEM client to make a database-style query request. The pseudocode for the operation is quite simple but hides a wealth of discussion and implementation issues for the WBEM server:

```
<object>*ExecQuery (
    [IN] string QueryLanguage,
    [IN] string Query)
```

This call says, "Here is a query which I have encoded in the X query language." This is, of course, of little use unless the WBEM server can interpret the X query language and so the DMTF has thoughtfully provided a discovery mechanism, which I describe on page 249, to allow the WBEM client to determine which query languages the WBEM server supports.

DMTF specification DSP0200 makes no recommendation about particular query languages. In practice, two query languages are emerging, WQL and CQL, as described on page 158.

Extrinsic Methods

Extrinsic operations are operations carried out by a method provider which may do anything. They may, for example, shut a system down, bring it up or perform any other complex operation.

In management systems such as SNMP, it is not possible to specify and invoke a particular function on a managed object—the equivalent functionality is achieved by having a side effect associated with a property; change this property and that function is automatically performed. This is less intuitive than simply calling a method and more liable to inadvertent invocation by the unwary programmer or operator. WBEM therefore provides a very simple way of remotely invoking a function (method) on an instance of a class.

As an example, consider the `PositionAtRecord()` function on the CIM_MessageLog class, as defined in the System common model. The *mof* for this class is shown in Figure 7.9 and, as the description of the class says, a CIM_MessageLog is any form of logging device.

Once stripped of all of the descriptions, this fragment of the CIM_MessageLog *mof* defines a method, called PositionAtRecord(), which, very informally, has the following C++-style prototype:

```
uint32 PositionAtRecord(char *IterationIdentifier,
                        bool MoveAbsolute,
                        long long *RecordNumber);
```

```
[Version ("2.7.0"), Description (
    "MessageLog represents any type of event, error or "
    "informational register or chronicle. The object "
    "describes the existence of the log and its "
    "characteristics. Several methods are defined for "
    "retrieving, writing and deleting log entries, and "
    "maintaining the log.") ]
class CIM_MessageLog : CIM_EnabledLogicalElement {

//  ==== many properties deleted ======

    [Description (
        "Requests that the Log's iteration identifier be "
        "advanced or retreated a specific number of records, "
        "or set to the entry at a specified numeric location. "
        "=== much description deleted === ") ]
    uint32 PositionAtRecord (
        [IN, OUT]
        string IterationIdentifier,
        [IN, Description ("Advancing or retreating the "
            "IterationIdentifier is achieved by setting the "
            "MoveAbsolute boolean to FALSE, and specifying "
            "the number of entries to advance or retreat "
            "as positive or negative values in the RecordNumber "
            "parameter. Moving to a specific record number is "
            "accomplished by setting the MoveAbsolute parameter "
            "to TRUE, and placing the record number into the "
            "RecordNumber parameter.") ]
        boolean MoveAbsolute,
        [IN, OUT]
        sint64 RecordNumber);
    };
```

Figure 7.9 *mof* for the PositionAtRecord Method

Note in passing the use of the IN and OUT qualifiers. These specify respectively that the parameter is an input to the `PositionAtRecord` function or is a returned (output) value. IterationIndentifier and RecordNumber are both: they are used to pass a value to `PositionAtRecord` and also bring a value back to the WBEM client.

The XML which the WBEM client would have to generate to invoke this function is listed in Figure 7.10. The namespace, name of the function, and values of the parameter are specified together with the key values of the instance to which the call is addressed. These allow the WBEM server to identify the instance and invoke a method with the given parameters. The PositionAtRecord function returns a uint32 value representing the success (zero) or failure (nonzero) of the method. The XML which would be returned by the CIM server to the client following successful completion is shown in Figure 7.11.

Most of the error codes which may be returned by the WBEM server following the invocation of an extrinsic method are the same as those described above for `EnumerateInstances`, but the new possibilities are:

- CIM_ERR_METHOD_NOT_FOUND to indicate that the extrinsic method does not exist.
- CIM_ERR_METHOD_NOT_AVAILABLE to indicate that, although the method exists, the CIM server was, for some reason, unable to invoke it.

Notice that, because of the query/response nature of the HTTP protocol, these extrinsic function invocations are effectively synchronous (i.e., the calling program (the client in this instance) is suspended until the result arrives back). Some work is being carried out within the DMTF to allow the invocation of extrinsic functions to be asynchronous (call and forget) but this has not reached maturity yet.

Authentication

On any client/server interface, questions of authentication arise: the server wants to authenticate the client to ensure that it is authorised to do the operations it requests and the client wants to authenticate the server so that it feels comfortable in passing confidential information such as passwords to it and so that it can trust the answers to its queries.

The DMTF's common models contain an infrastructure for managing authentication information for any user—see page 117. This section

```
<?xml version="1.0" encoding="utf-8" ?>
<CIM CIMVERSION="2.0" DTDVERSION="2.0">
 <MESSAGE ID="87872" PROTOCOLVERSION="1.0">
  <SIMPLEREQ>
   <METHODCALL NAME="PositionAtRecord">
    <LOCALINSTANCEPATH>
     <LOCALNAMESPACEPATH>
      <NAMESPACE NAME="root"/>
      <NAMESPACE NAME="cimv2"/>
     </LOCALNAMESPACEPATH>
     <INSTANCENAME CLASSNAME="CIM_MessageLog">
      <KEYBINDING NAME="CreationClassName">
       <KEYVALUE VALUETYPE ="string">ACNEComputing</KEYVALUE>
      </KEYBINDING>
      <KEYBINDING NAME="Name">
       <KEYVALUE VALUETYPE ="string">PrivateLog</KEYVALUE>
      </KEYBINDING>
     </INSTANCENAME>
    </LOCALINSTANCEPATH>
    <PARAMVALUE NAME="IterationIdentifier">
     <VALUE>ZZFRED762</VALUE>
    </PARAMVALUE>
    <PARAMVALUE NAME="MoveAbsolute">
     <VALUE>true</VALUE>
    </PARAMVALUE>
    <PARAMVALUE NAME="RecordNumber">
     <VALUE>145</VALUE>
    </PARAMVALUE>
   </METHODCALL>
  </SIMPLEREQ>
 </MESSAGE>
</CIM>
```

Figure 7.10   XML for Invoking the PositionAtRecord Method

```
<?xml version="1.0" encoding="utf-8" ?>
<CIM CIMVERSION="2.0" DTDVERSION="2.0">
 <MESSAGE ID="87872" PROTOCOLVERSION="1.0">
  <SIMPLERSP>
   <METHODRESPONSE NAME="SetPowerState">
    <RETURNVALUE>
     <VALUE>0</VALUE>
    </RETURNVALUE>
    <PARAMVALUE NAME="IterationIdentifier">
     <VALUE>ZZHFGD653</VALUE>
    </PARAMVALUE>
    <PARAMVALUE NAME="RecordNumber">
     <VALUE>145</VALUE>
    </PARAMVALUE>
   </METHODRESPONSE>
  </SIMPLERSP>
 </MESSAGE>
</CIM>
```

Figure 7.11   XML Returned by PositionAtRecord Method

deals more with the practical details of authenticating clients accessing the WBEM server.

The basic WBEM architecture, as illustrated in Figure 4.4 on page 39, contains a position for a security plug-in and this provides the focus for client authentication: it is really no more than an exit point to external authentication tools such as an LDAP or Radius server.

The normal mechanism for a WBEM client to request an operation from a WBEM server is, of course, through CIM-XML over HTTP. HTTP, as defined in RFC2068, uses a challenge/response authentication protocol: the client makes a request, the server offers the client a number of mechanisms by which it can authenticate itself, the client selects the most secure one that it supports and authenticates itself. The server then checks the authentication and, if satisfied, answers the request. This process is repeated for each operation carried out over HTTP, as illustrated in Figure 7.12.

The DMTF's "CIM Operations over HTTP" specification (DSP0200) specifies two types of authentication:

1. Basic Authentication, as described in RFC2068. This is recognised as being insecure, as passwords and user names are passed from the client to the server in plain text and are therefore sub-

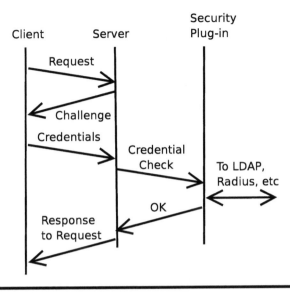

Figure 7.12   HTTP Authentication Exchange

ject to interception by a so-called "man in the middle" (MITM)
attack. Because of this vulnerability, DSP0200 forbids the use
of Basic Authentication in other than a secure environment unless it is combined with encryption of the type employed by the
Secure Sockets Layer (SSL).

2. Digest Authentication, as described in RFC1945 and refined in
RFC2617.

There are therefore three ways of authenticating a WBEM client:
Basic Authentication over SSL, Digest Authentication, or some other
protocol applicable to your application.

Different CIM server implementations support different authentication mechanisms: openPegasus, for example, supports only Basic Authentication over SSL whereas openWBEM supports both Basic and
Digest Authentication.

International Support

There are several issues associated with device and service management
systems which cross international borders, including:

- Providing support for international character sets in descriptions, class names, etc. defining the model. May I, for example, define a class called çáåùæü?
- Providing support for international character sets in error and other messages originating from the WBEM server or providers. How does the code of a provider know whether to return an error message in Greek, Russian, or English?
- Allowing an operator to specify locale (which may govern the model—device X is not available in country Y and therefore should not appear in menus) and language (which need not correspond to the locale).
- Allowing a string property to have several values simultaneously, each in a different language.

The issue of international support is covered briefly in Section 4.8 of the "CIM Operations over HTTP" specification (DSP0200). I address a variety of the international topics in the subsections below, but be aware that there can be significant differences between what the standard requires and what particular implementations currently deliver.

International Characters in CIM

·In *mof* specifications, a different approach has been taken to international alphabets depending on whether or not the results are likely to be read outside the programming community (which is expected to be able to read English).

It is assumed that there is no need for internal keywords (e.g., the keyword "class" in *mof* or the IMETHODCALL tag in a CIM-XML message) to be in other than English as spoken in the United States of America because these keywords are only seen by programmers.

For the names of classes, properties, and other names which might be seen by an operator, the DMTF's Common Information Model Specification (DSP0004) defines the subset of Unicode characters that may be used. This includes most international characters.

Languages between WBEM Client and WBEM Server

It is fortunate that the WBEM client/server interface is encoded in XML because XML already has support for tagging and encoding different character sets and languages: all XML processors are required to read UTF-8 encoded strings (UTF-8 is defined in RFC2279 and encodes characters using sequences of 1 to 6 octets—see the glossary entry for

Unicode). XML also allows the language of the contents of a message to be specified.

CIM-XML messages are, of course, commonly carried by HTTP(S) and HTTP supports support header fields known as Accept-Language and Content-Language. Accept-Language allows a client, when it invokes an exchange with a server, to specify the natural languages in which it can accept the response. For example,

```
Accept-Language: da, en-gb;q=0.8, en;q=0.7
```

means: "I prefer Danish, but will accept British English (with 80 percent acceptability) and other types of English (70 percent acceptability)."

Content-Language is set by both the client and the server and specifies the target audience for the contents of the HTTP message (not all of the words used in the message need be in the specified language—a page called "Teach Yourself French" is clearly aimed at an English-speaking audience, but will contain words in French).

International Providers

The XML and HTTP support described above is, of course, necessary, but not sufficient for a fully international interface between the WBEM client and server. In addition, the WBEM server has to exploit this underlying support and pass it on to the providers. You must therefore write your providers in such a way that they can exploit the information from the WBEM server. In particular:

- Ensure that all the strings in your code accept UTF-8 characters. openPegasus, for example, has a String class available for use by providers which stores the string using UTF-8 characters.
- Ensure that all constant strings (error messages, etc.) are included in a configuration file rather than being hard-coded into programs so that different languages can be accepted—handle Accept-Language from the server. Ensure that you rephrase messages which might be offensive in some cultures and which will not translate well ("Parent terminated and child killed").
- Ensure that you do not build messages from component parts. This may work in one language but lead to incorrect grammar in another.
- Avoid non-ASCII encodings (e.g., EBCDIC) if at all possible since ASCII maps cleanly to UTF-8 and other encodings.

Frequently Asked Questions

FAQ 24 *Table 7.1 on page 129, shows no methods for creating, deleting, and otherwise manipulating namespaces. Why?*

Namespaces are manipulated by using the normal intrinsic methods on the CIM_Namespace class.

FAQ 25 *What is the difference between a method such as EnumerateClassNames() and EnumerateClasses()?*

The first of these methods, eponymously, returns a list of the names of the classes. The second returns all details of the classes. Similarly for EnumerateInstanceNames() and EnumerateInstances()—the former simply returns a list of names, the latter all the gory details of the instance.

If a large number of instances is likely to be selected by a call to EnumerateInstances(), it may be better to call EnumerateInstanceNames() and then call GetInstance() on each returned name as this will spread the traffic load.

FAQ 26 *Where can I find additional examples of the XML which is sent between the WBEM client and the WBEM server?*

In the appendices of standard DSP0200. Examples of the XML requests and responses are given for many of the major operations.

FAQ 27 *Figure 7.7 shows the XML sent to the WBEM server to create an instance of CIM_IndicationFilter. When I compare it with Figure 7.8 I notice that it does not have KEYBINDINGS. Why?*

The entity which is passed to DeleteInstance is an instance name (object path). The entity which is passed to CreateInstance is an instance (and an instance name is returned). This is another manifestation of the name/instance difference discussed in FAQ 25.

When deleting an instance, the WBEM server only needs the name (path) of the instance. This name is simply the concatenation of the key properties.

When creating an instance, the WBEM server needs all of the properties of the instance, not just the keys. It has access to the definition of the class and can therefore deduce which of the properties are keys.

FAQ 28 *Does the WBEM client/server interface support a transactional model whereby a number of configuration changes can be entered and committed (or abandoned) together?*

No. A presentation on this topic was given by Cisco Systems at the DMTF 2003 Global Management Conference. The presentation outlined a number of possible approaches, none of which has yet been accepted.

It is this lack of transactional support which primarily gives WBEM its enterprise rather than carrier flavour. In the high reliability world of the carrier it is not acceptable that every command be entered and executed individually, perhaps placing the system temporarily into an inconsistent state. It must be possible to enter batches of commands, have the anticipated results checked for consistency before they are applied and then have them executed atomically: all the commands or none being executed. Similarly, it must be possible for the system to roll back a transaction to remove the effect of a set of commands.

The need for a transactional model is recognised throughout the industry: RFC3512, for example, when discussing SNMP management contains the paragraph:

> Configuration activity causes one or more state changes in an element. While it often takes an arbitrary number of commands and amount of data to make up configuration change, it is critical that the configuration system treat the overall change operation atomically so that the number of states into which an element transitions is minimized. The goal is for a change request either to be completely executed or not at all. This is called transactional integrity. Transactional integrity makes it possible to develop reliable configuration systems that can invoke transactions and keep track of an element's overall state and work in the presence of error states.

Transactions are already present to some extent within WBEM since a service can be modelled in CIM. This allows an operator to enter a single command to create an instance of the service. A provider for that service could then execute a number of lower-level commands to provision the individual components of the service. Such a provider could also have the knowledge necessary to roll back (i.e., remove) the actions it has already taken if a later action fails. This gives the impression of a transaction-oriented interface to the operator but the responsibility for the transaction lies with each individual provider, with little help from the WBEM architecture.

FAQ 29 *What is the difference between HTTP and HTTPS?*

HTTP (Hypertext Transfer Protocol) is a stateless protocol used between a client (making a request) and a server (responding to the request). Each exchange consists of a message going from the client to the server and a response message coming back—no state is held at the server between exchanges. This is the protocol which browsers use across the Internet to fetch information from servers.

HTTPS (Secure Hypertext Transfer Protocol) is HTTP over Secure Sockets Layer (SSL), which means that the HTTP packets are encrypted before being sent.

Chapter 8

The Listener Interface

This chapter relies heavily on the Event Common Model and, at the time of writing, several parts of this model are undergoing modification between version 2.7 and 2.8. I have tried to include the most up-to-date class hierarchy but, since some of this is experimental at the time of writing, it may change before version 2.8 finally appears.

The Indication Mechanism

Events occur and conditions causing alarms arise spontaneously in all systems. Some of these need to be brought to an operator's attention, some need to be logged for later review, some need to be passed to monitoring and analysis equipment (the "P" for performance in FCAPS) and many can be ignored. The handling of a particular alarm may change from minute to minute: perhaps an operator is trying to pin down an elusive fault and wants to be alerted as soon as it occurs, perhaps a particular fault is only important on certain days or if another fault has not also recently occurred.

The WBEM server provides the interfaces to allow operators to specify precisely the faults for which they want to be notified and to turn the notification on and off. The classes used to establish this monitoring are included in the Event Common Model.

There are several components in the path between an event occurring in a piece of hardware or software and an indication arriving on the screen of an operator. These include:

The Indication Provider which detects the alarm and passes it to the WBEM server as an indication. An indication is an instance of a class derived from CIM_Indication.

The Filters which the WBEM server consults to see whether the indication satisfies a predefined pattern. Filters are instances of classes derived from CIM_IndicationFilter. A filter might, for example, select all events arising from the class `MyTemperatureAlarm` where the

```
Severity == Critical AND Temperature is > 180
```

The Subscriptions which the WBEM server consults to see whether any operator is interested in being notified about the indication. Subscriptions are instances of associations derived from CIM_IndicationSubscriptions.

The Handlers to which the WBEM server sends the notification. Handlers are defined in instances of classes derived from CIM_IndicationHandler.

The Listeners which represent the operators and receive the indications on their behalf.

The Consumers, the CIM-speak term for the operator's software which will presumably display the indication on a screen, write it to a file, etc.

This whole mechanism is illustrated in Figure 8.1 where time is assumed to flow from top to bottom of the picture. The flow of activity in the figure is as follows:

1. An operator (or other WBEM client such as a *mof* compiler for static instances) creates an instance of the class representing the handler.
2. The operator creates an instance of the filter class describing precisely the types of indication in which he or she is interested.
3. The operator creates an instance of an association linking the filter to the handler.
4. When an event occurs, the associated provider (an Indication Provider) creates an instance of the Indication.
5. The WBEM server is aware of this instantiation and examines the filters to determine which filters allow it through. It then follows the associations to see which handlers (if any) are interested in this event.

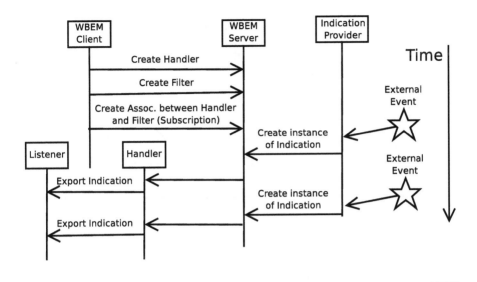

Figure 8.1 The Indication Mechanism

6. The WBEM server then passes selected information from the indication instance to the handler.
7. The handler then passes it to the listener for display.

Figure 8.2 gives a little more detail of how filters and handlers may be combined. On my toaster there is an alarm, called **TemperatureAlert**, which signals a high or low temperature on the heating element. I have created two filters: one passing (accepting) alarms if the temperature becomes too low, the other passing alarms if it becomes too high. Handler 1 sends a copy of the Indication to the local fire station which is interested in the temperature becoming too high whereas Handler 3 is connected to my toaster repair shop which is only interested in the temperature becoming too low. I am monitoring Handler 2 myself and need to know of either event so it receives both types of Indication. Note that Figure 8.2 is an informal diagram: it is not UML.

If you are creating a listener to receive information about a particular alarm then you need only create a handler for it, define a filter (unless one already exists) and then create an instance of the association (the subscription) between the filter and the handler. Your listener will then start to receive events from the managed device as they occur. What it does with them is, of course, outside the scope of WBEM/CIM but could include writing a log file, sending a message to a pager, opening a window on an operator's screen or sending an

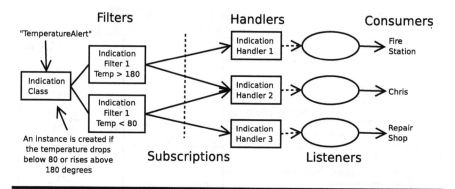

Figure 8.2 Indication Filters and Handlers

e-mail (or instant message) to an operator. If the operator is managing the device with a Web browser, then possibly a Java thread could be waiting on the arrival of these messages and creating a pop-up window when they arrive.

In summary, instances of four classes are required to make all of this work: CIM_Indication, CIM_IndicationFilter, CIM_IndicationHandler and CIM_IndicationSubscription. These are described in the following sections.

Indications

Figure 8.3 illustrates the top level of the Indication class hierarchy. CIM_Indication is a very generic class and is defined with the **ABSTRACT** qualifier meaning that instances of it cannot be created; it is simply a superclass for all indications. It has three subclasses:

1. CIM_ClassIndication, used to describe events arising from the manipulation of classes: their creation, deletion, and modification. In this case, the CIM server itself effectively acts as the Indication Provider, creating the Indication when a class is manipulated.
2. CIM_InstIndication, used to describe events arising from the manipulation of instances. As with CIM_ClassIndication, this includes their creation, deletion and modification but also includes the invocation of a method on them. Again, the WBEM server acts as the Indication Provider. A listener could be set up to receive indications related to class and instance manipulation by subscribing to CIM_InstIndications

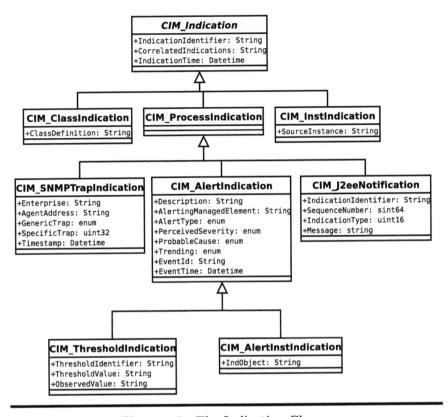

Figure 8.3   The Indication Classes

and CIM_ClassIndications and, perhaps, create a log file as part of a security trace.

3. CIM_ProcessIndication, used for all "external" events—events not raised because of the manipulation of the model. In principle, this type of indication is not actually required because all externally occurring events could be mapped to an operation in the model (creation of an instance, for example). In practice, most event handling does occur through CIM_ProcessIndication because this class forms a superset with most of the standard properties required for an alarm notification: see Figure 8.3. This type of Indication was designed to allow an alarm detected deep in an operating system or telecommunications stack and possibly not associated directly with any CIM object to be signalled as easily as possible.

The subclasses of CIM_ProcessIndication as illustrated in Figure 8.3 are:

SNMP	CIM	Description
Enterprise	Enterprise	Describes the type of entity generating the trap
Generic-Trap	GenericTrap	Describes the type of trap (e.g. authentication failure, link up, link down)
Specific-Trap	SpecificTrap	Identifies the precise trap
Time-Stamp	TimeStamp	Time at which the trap was generated relative to the last reset of the device

Table 8.1   Some SNMP to CIM_SNMPTrapIndication Mappings

- **CIM_SNMPTrapIndication.** This is a concrete (i.e., nonabstract) class to which SNMP traps* may be mapped. The class may be used as a simple encapsulation of the SNMP trap, in which case, in order to interpret it, the receiver will need access to the managed object's Management Information Base (MIB). To avoid the need for propagating the MIB, a better (but perhaps longer-term) solution might be to model the managed object in CIM. The fields from the SNMP trap are mapped into the properties of the CIM_SNMPTrapIndication: see Table 8.1 for some of the mappings.

- **CIM_AlertIndication.** This is also a concrete (i.e., nonabstract) class and is the preferred superclass for operational alarms and events arising other than from SNMP. Many of the properties of the **CIM_AlertIndication** mirror similar properties defined in standard X.733 (Systems Management: Alarm Reporting Function) from the International Telecommunication Union (ITU-T). Table 8.2 gives some of the relationships between X.733 and the properties of **CIM_AlertIndication**.

- **CIM_J2eeNotification.** This is a concrete class into which "Java 2 Enterprise Edition" (J2EE) alarms may be mapped. J2EE is an initiative by Sun Microsystems for the management of Web Services.

---

* A "trap" is SNMP-speak for an indication; see RFC1157.

X.733	CIM_AlertIndication	Note
Additional Text	Description	
Event Type	AlertType	Defines type of alert, e.g., security alert, environmental alert, processing error.
Perceived Severity	PerceivedSeverity	Defines importance of the alarm, e.g., Informational, Warning, Minor, Major, Critical, Fatal.
Probable Cause	ProbableCause	Origin of the alarm, e.g., Bandwidth reduction (showing the telecommunications origin of X.733), Connection Establishment Error, Application Subsystem Failure, Toxic Leak Detected.
TrendIndication	Trending	Estimate of whether this situation is getting worse or better.

Table 8.2   Some X.733 to CIM_AlertIndication Mappings

Indication Filters

Once the Indication classes have been defined as described above, it is necessary to create filters which allow particular types of indications to pass through.

Figure 7.7 on page 137 contains an example of the XML that a WBEM client could use to create a CIM_IndicationFilter. It would be useful for you to examine that example closely while reading this section as it shows the properties in an instance of CIM_IndicationFilter.

Figure 8.4 shows part of the *mof* definition of the CIM_IndicationFilter class (I have omitted many of the properties). The properties of greatest interest are:

Name. This is the name of the filter. Being able to identify a filter allows it to be reused in different contexts.

Query. This is the actual filter which defines whether or not a particular Indication should be passed through or be dropped. The format of the filter depends on the language chosen to express it (see QueryLanguage). An example, using the WQL, might read as follows:

```
SELECT Temperature, Name
```

```
FROM TemperatureIndication
WHERE Name != "Alison"
```

to accept all TemperatureIndication indications which do not refer to a toaster with the name "Alison." In particular this filter will extract the Name and Temperature properties for passing to the listener.

QueryLanguage. In principle, the Query Language may be any language comprehensible to the WBEM server. In practice, two languages are commonly used:

■ Web Query Language (WQL): A dialect of SQL as defined by ANSI with extensions to support CIM. A query in WQL might look like:

```
select * from Solaris_DiskDrive where
                     Storage_Capacity = 1000
```

■ CIM Query Language (CQL): this language is defined by the DMTF in document DSP0202 (Query Specification) and is also based on SQL and XML-Query as defined by W3C, the World Wide Web Consortium.

Handlers

The manner in which a handler is invoked by the WBEM server varies from implementation to implementation. Typically handlers will look like special types of providers and have a similar interface, supporting, for example, methods such as **handleIndication()** which accept the indication instance.

The major question is how the handler finds its associated listeners. In earlier versions of the Event Common Model, the class CIM_IndicationHandler was defined to hold information about particular handlers and their listener destinations. In version 2.8 of the model this class has been deprecated and replaced by CIM_ListenerDestination, of which CIM_IndicationHandler is now a subclass. This illustrates the manner in which classes are deprecated and replaced: CIM_IndicationHandler still exists in the model, and because it has been moved down the hierarchy from its previous position, all existing implementations will continue to work. The **DEPRECATED** qualifier, however, discourages new usage.

```
// ================================
// IndicationFilter
// ================================
[Version("2.6.0"), Description (
    "CIM_IndicationFilter defines the criteria for generating "
    "an Indication and what data should be returned in the "
    "Indication. It is derived from CIM_ManagedElement to "
    "allow modeling the dependency of the filter on a specific "
    "service.") ]
class CIM_IndicationFilter: CIM_ManagedElement
    {
    // Note: many properties deleted

    [Key, Description ("The name of the filter.") ]
    string Name;

    [Required,
        Description ("A query expression that defines the "
        "condition(s) under which Indications will be "
        "generated. For some Indication classes, the query "
        "expression may also define the instance properties "
        "to be copied to the CIM_InstIndication's "
        "SourceInstance and PreviousInstance properties. "
        "Query language semantics include projection "
        "(e.g., Select), range (e.g., From) and predicate "
        "(e.g., Where)."),
        ModelCorrespondence
        {"CIM_IndicationFilter.QueryLanguage"} ]
    string Query;

    [Required, Description (
        "The language in which the query is expressed.") ]
    string QueryLanguage;
    };
```

Figure 8.4 Part of CIM_IndicationFilter

The actual address (e.g., "localhost:5988/myHandler1") of the listener is not held in the CIM_ListenerDestination instance, but in a subclass thereof, the subclass depending on the protocol. At present only CIM_ListenerDestinationCIMXML is defined.

Typically, when the WBEM server invokes a handler it will pass it an instance of the indication and an instance of CIM_ListenerDestination. With these two pieces of information the handler can reformat the indication into export format and pass it to the listener.

Subscriptions

Having created the indication and filter classes, we now need to tie a particular filter to a particular handler by creating an instance of an association descended from CIM_IndicationSubscription. This will cause the WBEM server to inform the provider registered to generate the indications that someone is listening for them. If this is the first such subscription then this may cause the provider to start generating indications when the external events occur.

I list part of the *mof* for the CIM_IndicationSubscription class in Figure 8.5. This is a straightforward association between a filter and a handler, represented by a CIM_ListenerDestination.

Listeners

Once all of the infrastructure described above has been set up, with filters, subscriptions, and handlers in place, indications start to flow to the handlers. The handlers then initiate transfers to the listeners and, since the listeners cannot be assumed to have knowledge of CIM namespaces, etc., the format of the XML is known as CIM Export Format and is slightly different from what I have described elsewhere in this book. I include an example in Figure 8.6 and, if you are an astute XML reader, you should have no problem in decoding it. Effectively, the namespace has been removed and more or less everything else remains unchanged.

Although the export process is designed to convey information in one direction, from the handler to the listener, the definition of the export interface is sufficiently general to allow parameters to be passed back if it is used for other, currently undefined, applications.

```
// ================================================================
// IndicationSubscription
// ================================================================
[Association, Version ( "2.7.1000"), Description (
    "CIM_IndicationSubscription describes a flow of "
    "Indications. The flow is specified by the referenced "
    "Filter, and directed to the referenced destination or "
    "process in the Handler. Property values of the "
    "referenced CIM_IndicationFilter instance and "
    "CIM_ListenerDestination instance MAY significantly "
    "affect the definition of the subscription. E.g., a "
    "subscription associated with a \"Transient\" "
    "destination MAY be deleted when the destination "
    "terminates or is no longer available.") ]
class CIM_IndicationSubscription {

    [Key, Description (
    "The Filter that defines the criteria and data of the "
    "possible Indications of this subscription.") ]
    CIM_IndicationFilter REF Filter;

    [Key, Description (
    "The Handler addressing delivery of the possible "
    "Indications of this subscription.") ]
    CIM_ListenerDestination REF Handler;

    // many properties deleted
    };
```

Figure 8.5   Part of CIM_IndicationSubscription

Frequently Asked Questions

FAQ 30   *What is the difference between a listener and a consumer?*

A consumer is the application which actually handles an indication, possibly recording it, possibly displaying it to an operator. A listener is the destination to which an indication is sent by the WBEM server; the listener is known to the WBEM server, the consumer is not. Any number of consumers may be associated with a single listener.

FAQ 31   *Is the WBEM client which sets up the subscription the same as the listener which receives the Indications?*

```
<CIM CIMVERSION="2.0" DTDVERSION="2.0">
  <MESSAGE ID="1007" PROTOCOLVERSION="1.0">
   <SIMPLEEXPREQ>
     <EXPMETHODCALL NAME="ExportIndication">
      <EXPPARAMVALUE NAME="NewIndication">
       <INSTANCE CLASSNAME="CIM_AlertIndication" >
         <PROPERTY NAME="Description" TYPE="string">
           <VALUE>Sample CIM_AlertIndication indication</VALUE>
         </PROPERTY>
         <PROPERTY NAME="AlertType" TYPE="uint16">
           <VALUE>1</VALUE>
         </PROPERTY>
         <PROPERTY NAME="PerceivedSeverity" TYPE="uint16">
           <VALUE>3</VALUE>
         </PROPERTY>
         <PROPERTY NAME="ProbableCause" TYPE="uint16">
           <VALUE>2</VALUE>
         </PROPERTY>
         <PROPERTY NAME="IndicationTime" TYPE="datetime">
           <VALUE>20010515104354.000000:000</VALUE>
         </PROPERTY>
       </INSTANCE>
      </EXPPARAMVALUE>
     </EXPMETHODCALL>
   </SIMPLEEXPREQ>
  </MESSAGE>
</CIM>
```

Figure 8.6   Example of Export Format

No. There are two distinct processes at work here—the client uses the normal WBEM client/WBEM server interface to create the subscription on behalf of the CIM listener. The CIM listener then receives the indications. Although the two functions are distinct, there is, of course, no reason why they should not be implemented in the same software or in two software modules running on the same workstation.

FAQ 32   *In what order are Indications delivered to listeners?*

They may be delivered in any order; that event A occurred before event B or that Indication A' was created before Indication B' does not imply in any way that A' will be delivered before B'.

FAQ 33 *If the WBEM server fails and recovers, will my subscriptions still be there?*

The persistence of subscriptions (or instances of any class) across a failure is not discussed in the specification—assume the worst, that they are gone and need to be recreated.

PRACTICE

Chapter 9

Building Your Own Model

The PBX Example

Modelling is an art rather than a science. All that I can give here are a few commercial and technical tips on how to approach it.

In order to make the exercise more concrete, I decided to use a simple device as an example but selecting a suitable device was not easy—I looked for something for which the DMTF did not already have a model (thereby ruling out almost everything to do with computers or IP routers), something which all readers would readily understand (thereby ruling out the more esoteric telecommunications devices), something that was reasonably complex and provided services (thereby, unfortunately, ruling out my toaster) and something that was sufficiently uncomplicated that the model could be largely complete.

In the end I settled on a small, primitive, and sparsely featured PBX, the telephone switching device used within a company or hotel to route calls between extension telephones and between extensions and outside lines. Figure 9.1 illustrates the main features of the device:

- It can hold up to eight telephone interface cards, each of which provides connectors for eight telephones.
- It can hold up to four trunk interface cards, each of which provides two connections to the public telephone system.
- It contains a microprocessor to control the various switching devices. This microprocessor runs Linux and the Bayonne open source PBX software (see `http://www.gnu.org/software/bayonne/`) which it loads from a hard disk built onto the processor card.

167

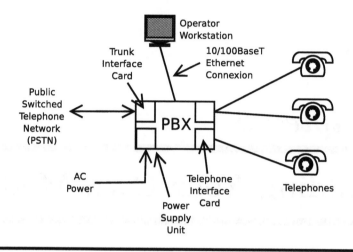

Figure 9.1   PBX Interfaces

■ It is powered from the AC electricity supply.

A knowledge of the PBX's physical elements will also be of importance to the modelling work. I assume that it is shipped as a small shelf into which various cards may be plugged; see Figure 9.2. The power supply unit, processor card, and at least one telephone interface card need to be present to make a working system. Additional telephone interface cards and trunk interface cards can be added as required.

Telephony statistics are collected by the processor (number of calls made from each telephone, number of incoming calls blocked by busy telephone, number of outgoing calls blocked by insufficient trunks, etc.) and these need to be made available to an operator.

Maintenance Operators are not allowed access to the telephony statistics but are allowed access to information about the operating system (memory usage, time since last reset, disk usage, etc.) and may run tests on all cards.

Tests are additionally run periodically on all the software and hardware in the PBX and alarms are generated if any problems are discovered.

This "specification" covers only a very small part of the definition of a PBX and leaves more questions unanswered than it answers: what are the precise authority levels of each type of user, how are the users to be authenticated, are the cooling fans and other hardware components to be monitored, what alarms arise spontaneously from the system, etc.? I believe that building a model of the components I have defined will

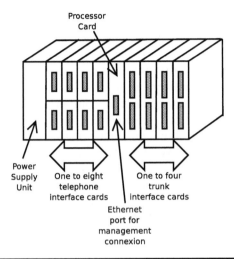

Processor
Card

Power
Supply
Unit

One to eight
telephone
interface cards

One to four
trunk
interface cards

Ethernet
port for
management
connexion

Figure 9.2 PBX Field-Replaceable Units

give you sufficient confidence to extend the models as needed without extending this book too much with replication.

Commercial Decisions

Assuming that your company is manufacturing, and hoping to sell, this PBX, one decision you will have to make is how to handle the model that you are going to create—will you try to initiate or at least encourage a standardisation process through the DMTF, will you publish the model so that manufacturers of management systems can access it or will you keep the model secret as an asset for your company?

With a simple device like this PBX, this is probably not an important decision—there are few *mof* classes and their positions within the common and core models are fairly obvious—but if your company were to manufacture a genuine PBX, a core router or piece of optical transmission equipment then the decision would be harder. You would need to address some or all of these questions:

■ Are you dominant in your market and, if so, are there small companies trying to break in who might club together and use a standardised management strategy as a lever against you ("buy from us and get standard management interfaces, buy from our large competitor and get locked into a proprietary one")? Or are you one of the small companies trying to break in?

■ Can you afford the time to take your model through the standardisation process? Before the DMTF will be willing to accept your model, at least two companies will need to present it together (to ensure interoperability). Are you willing to work with one of your competitors to do this?

■ Is the ease of management a competitive advantage for your product? Note that this does not indicate whether or not you have a wonderful management workstation with excellent graphical user interface. WBEM/CIM does not address the operator interface—only the storage and manipulation of the on-device management information.

■ Do you make money from selling the management workstations or just by selling the PBXs? If you make money from the workstations then making the model public is probably going to hit your sales. If, on the other hand, you only make real money from selling the PBXs, then a PBX which can be managed by an industry-standard management system might be more attractive to your customers.

■ Will your PBX be installed alongside other devices (IP routers, storage arrays) which are already managed by WBEM/CIM? If so, could you gain competitive advantage by allowing your customers to manage your device in the same way?

Deciding What to Model

We only need to model the items that we wish to manage so it is important to list those items before we create the model. I am going to assume that, in no particular order, we will want to manage the following items:

■ The PBX itself, at least to disable and enable it.

■ The telephone and trunk interfaces, to find how many are fitted, whether they are enabled or disabled, and to get statistics about their usage.

■ The processor card, to find out about the amount of disk space remaining, the status and statistics of the Ethernet port, etc. Note that we must prevent the operator from disabling the Ethernet port because this will break his or her own management connection and make the PBX unmanageable.

■ The operating system and application software running on the processor card.

- The diagnostic tests, to invoke them manually and to retrieve results.
- The telephony service, to create groups of telephones for assigning incoming calls and to enable or disable the access of particular telephones at particular times of the day to outgoing trunk lines. The basic telephony statistics are also included in this service.
- The voice mail service, to enable and disable access to voice mail for particular telephones and set individual limits on the disk space available for each voice mail user.
- The users, to set passwords, etc.

Modelling Guidelines

Although they were written for a very different modelling paradigm, SNMP, most of the modelling guidelines contained in RFC1493 and enhanced in RFC3512 are relevant to CIM modelling:

- Start with a small set of essential objects and add only as further objects are needed. Minimising the total number of objects is not an explicit goal, but usability is. Be sure to consider deployment and usability requirements.
- Require objects be essential for either fault or configuration management.
- Consider evidence of current use and utility (i.e., look to see how the parts of the model you are considering extending are already used).
- Generally exclude objects which are simply derivable from others, but consider the impact on a management application. If an object can help reduce a management application's complexity, consider defining objects that can be derived.
- Avoid causing critical sections to be heavily instrumented (i.e., do not add lots of statistical counters to a piece of code for which high-performance is required).

In the modelling of the PBX example below, I try to follow these guidelines.

Constraints on Our Models

Because no common model exists which is adequate for our PBX device we will have to create an extension schema. This will inherit from the core and common models but will be specific to our device. Before we launch into our schema, it is worth considering the ways in which we may modify and extend the core and common models. We could:

- Add (or delete) a property to (from) an existing class or subclass
- Add (or delete) classes
- Turn an existing class into an Association
- Add or delete qualifiers
- Add (or delete) a method to (from) a class

Some of these activities will cause problems to other management systems. Deleting the CIM_OperatingSystem class, for example, or moving it other than downwards in its own hierarchy may not matter for your management system because it will know where to find it. If, however, your customer decides to manage your PBX with the same management system that he is using for your competitor's product, then it may not expect the class other than in CIM_EnabledLogical-Element's subtree.

We will therefore try to avoid deleting properties or classes from the core and common models.

Naming the Schema

Another decision we need to make is the name for our schema (model). As you have seen, all of the core and common models belong to the CIM schema and so all class names begin with CIM_. An obvious, and bad, first choice for schema name might be PBX. Even if our intention is never to release our schema into the public domain, the possibility of a name clash would arise if some other company started issuing models with the PBX schema name. In order to avoid this, the DMTF gives the following advice:

> "CIM" and "PRS" are reserved schema names and MUST NOT be used by any company/organization other than the DMTF.
>
> The schema name MUST be unique and MUST begin with an alphabetic character. We recommend the use of one of the following methods to assure uniqueness:

- Use a trademark registered by your company eliminating any "dot" characters from the trademark, and use the result as your schema name.
- Make use of your company or organization's registered DNS entry:
    - Use the right-most elements of your company or organization registered DNS address up to and including your company or organization name.
    - Move any elements to the right of your company or organization name to the left of that name.
    - Eliminate all "dot" characters.
    - Use the result as a prefix to your schema name.
- Use another unique name that you can be sure is unique.

Assuming that we work for the Acne Manufacturing Company, a name which we will assume for this example has been trademarked, a suitable schema name would be ACNE and our classes would have names of the form ACNE_classname. However, Acne manufactures devices other than PBXs and would like to differentiate the PBX product where necessary. There is some confusion in the DMTF's specifications about whether the class name itself may contain an underscore (see the discussion on page 192) and, to avoid possible problems, I will not use additional underscores and will use class names in the format ACNE_PBXxxxxx where xxxxx represents the rest of the class name.

Note that this may be a poor choice for various reasons: perhaps Acne uses some of the PBX's components and software on another, non-PBX, product. In this case we would like to reuse the classes that we are going to create and the inclusion of PBX in the class name would be unfortunate.

Positioning the PBX Class

Warning! *I walk through the selection of this first class in excruciating detail, leaving bread crumbs behind, showing my train of thought. Don't despair; once through this first selection, I speed up substantially.*

The fundamental problem in creating an extension schema is knowing where to start. There are over 900 classes in version 2.8 of the core and common models and we need to find the classes from which we should inherit. It is essential that we choose carefully because, if we choose wrongly, our choice will be difficult to change later. The basic technique that we will employ is to keep coming back to the basic inheritance rule, *is a*, and ask ourselves repeatedly, "is it true that my class *is a* X?" Obviously the descriptions given in the core and common class definitions are a great help in this process.

Even when we have found what we believe are the correct locations for our classes, it may be necessary to relocate them (or add additional associations) for performance reasons.

So, we will start at the top and assume that we are going to define a ACNE_PBX class to model the entire PBX. This is presumably going to be a CIM_ManagedElement (as everything is). Looking at Figure 6.1 on page 91 (or, better still, at the full UML diagrams issued by the DMTF for the latest release of the core and common models), we may arrive at the CIM_System and the CIM_LogicalDevice classes as possible representations for our PBX. Certainly a PBX is, in the common usage of the term, a "system" and it is also a "device."

On examining the full model, the subclasses of CIM_LogicalDevice appear to be a bit low-level for us: power supply, cooling device, door, power supply, etc. Our PBX certainly includes these but is an altogether larger entity. Figure 6.1 also shows a useful aggregation: CIM_SystemDevice which combines logical devices into a system—more evidence that our PBX *is a* system.

I give a short description of the System Model starting on page 99 of this book and the definition written into the DMTF's *mof* file contains:

> *CIM_System represents an entity made up of component parts (defined by the SystemComponent relationship*) that operates as a "functional whole." It should be reasonable to uniquely name and manage a System at an enterprise level. For example, a ComputerSystem is a kind of System that can be uniquely named and independently managed in an enterprise. However, this is not true for the power supply (or the power supply subsystem) within the computer.*

Certainly our PBX is an entity made up of component parts (chassis, boards, power supply, application software, operating system, etc.) and it operates as a functional whole. The CIM_System class seems to be a good place to start, but before we declare our PBX class to be a

---

* Beware: Only use the CIM_SystemComponent association if no lower-level association, such as CIM_SystemDevice, exists and will do the job. CIM_SystemDevice *is a* CIM_SystemComponent.

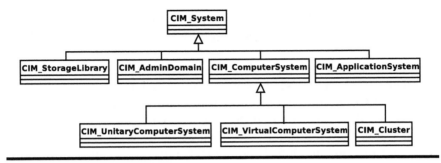

Figure 9.3   CIM_System's Children (and Grandchildren)

subclass of CIM_System, we should look at its existing subclasses to see whether any more refined (specialised) class fits even better.

In version 2.8 of the common models, there are only four subclasses defined for CIM_System (see Figure 9.3; CIM_ApplicationSystem, CIM_AdminDomain, CIM_StorageLibrary, and CIM_ComputerSystem). Our PBX is certainly neither an Admin Domain nor a Storage Library but we need to examine Application System and Computer System more closely.

The description in the *mof* file defining CIM_ApplicationSystem says:

> *The CIM_ApplicationSystem class is used to represent an application or a software system that supports a particular business function and that can be managed as an independent unit. Such a system can be decomposed into its functional components using the CIM_SoftwareFeature class. The Features for a particular application or software system are located using the CIM_ApplicationSystemSoftwareFeature association.*

This sounds a bit too software-oriented for us. CIM_ComputerSystem, on the other hand, is described as:

> *A class derived from System that is a special collection of ManagedSystemElements. This collection provides compute capabilities and serves as aggregation point to associate one or more of the following elements: FileSystem, OperatingSystem, Processor and Memory (Volatile and/or NonVolatile Storage).*

This sounds much closer to our PBX: it certainly contains computing capabilities and it serves as an aggregation point for several of the stated elements.

Once again we should look at the available subclasses and check whether any of them describe our system more completely—see Figure 9.3. CIM_ComputerSystem's subclasses are CIM_UnitaryComputer-System, CIM_VirtualComputerSystem and CIM_Cluster.

Of these, CIM_Cluster is described as being for a system "that 'is made up of' two or more ComputerSystems which operate together as an atomic, functional whole to increase the performance, resources and/or Reliability, Availability, and Serviceability."

While we may one day add a second processor to our PBX, it will not be primarily to increase performance or availability; so, we will avoid CIM_Cluster.

CIM_UnitaryComputerSystem, on the other hand, is described as representing "a Desktop, Mobile, NetPC, Server or other type of a single node Computer System." This is more tempting, but its use would preclude the introduction of a second processor (perhaps on a second processor board) at a later date.

A useful sanity check on the positioning of ACNE_PBX is to look at the properties and associations that it has inherited and determine inhowfar they relate to our PBX. If we found them to be totally irrelevant to a PBX then this might make us suspicious that we had chosen wrongly. Some of the properties certainly seem to be relevant:

Dedicated: Describes whether this is a general-purpose computer or one dedicated to a particular function. This is an enumerated variable and its values, although not containing PBX (unfortunately), do contain devices which are not dissimilar to PBXs in their own field: router, switch, layer 3 switch, central office switch, hub, access server, firewall, file server, repeater, bridge/extender, gateway.

ResetCapability: Defines whether or not the PBX's computer can be reset by a hardware switch.

PrimaryOwnerName: Defines the owner of the PBX.

CIM_InstalledOS and **CIM_RunningOS:** These associations link our PBX with its operating system.

CIM_HostedService: This association allows us to tie our PBX to the services it offers.

CIM_SystemPackaging: This association ties our logical PBX to its physical packaging.

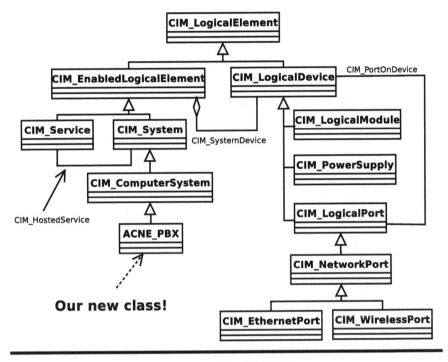

Figure 9.4   The PBX Class

CIM_SystemDevice: As described above, this allows us to tie the lower-level devices (power supplies, etc.) together into our PBX system.

These sound reasonable and so I will assume that the PBX *is a* CIM_ComputerSystem and that its component parts are CIM_LogicalDevices; see Figure 9.4.

Modelling the PBX's Components

Note: The classes which I decide to add into our extension schema are listed for convenience in Table 9.1 on page 185.

Having added the PBX class itself, we still need to model the devices which comprise it (the telephone and trunk ports, the interface cards, processor, disk drive, power supply, etc.), the grouping of telephones

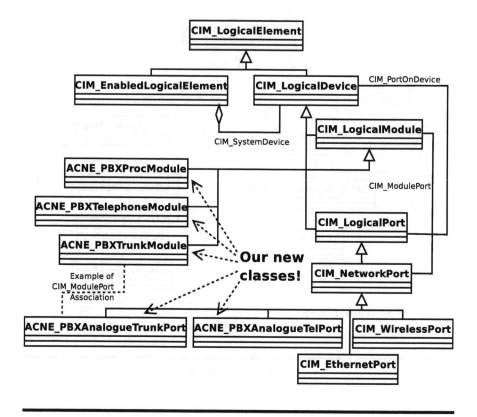

Figure 9.5   The PBX Device Classes

and the allocation of incoming trunks to the groups, the users and the services the PBX provides.

Luckily most of these devices are already modelled and I have included part of the logical device subtree in Figure 9.4 for convenience but our PBX has a number of port types which do not yet appear in the common models. One way to find a home for them is to realise that IP routers and other network equipment also have ports, so a class for ports is likely to exist somewhere; a little searching leads to CIM_TokenRingPort, CIM_EthernetPort and other similar ports living as subclasses of CIM_NetworkPort. Even though many of the properties of CIM_NetworkPort are oriented towards data rather than telephony ports (e.g., bitrate), it would seem that our ports are peers of these and so, as shown in Figure 9.5, we can add two additional types of port there. A nice feature of this is that the CIM_SystemDevice association already exists to link these to our PBX class.

The other port on our PBX, the Ethernet through which we manage it, is already well modelled so we have no work to do there. I have also shown these port classes in Figure 9.5.

The other items we need to model are the interface cards. Remember from the discussion on page 89 that the interface cards may be both physical items with a serial number and physical location, and also a logical item describing the function performed. In the remainder of this chapter, I will deal exclusively with the logical function of the interfaces. This does not mean that the physical characteristics are unimportant or should not be modelled—simply that the physical models are relatively straightforward. If it is important to manage the hardware inventory of the PBX then the CIM_PhysicalElement, CIM_FRU and CIM_Product classes could be subclassed and exploited. As they say in mathematical textbooks, this is left as an exercise for the reader.

The logical view of the interface functions fit as subclasses of CIM_LogicalModule which is described in its *mof* as "the logical device corresponding to a line card/blade in a device. For example, a line card in a switch is an instance of LogicalModule, associated with the switch itself. A logical module is not necessarily independently managed." This seems to fit our requirements nicely and there is even a predefined association CIM_ModulePort which will allow us to associate each module with the port types which it supports. I therefore introduce the three classes ACNE_PBXProcModule, ACNE_PBXTelephoneModule, and ACNE_PBXTrunkModule.

Modelling the Statistics

The PBX collects various statistics about the number of calls made by each telephone, the number of times an outgoing call has been blocked by all the trunks being busy, etc. We need to model some form of container to collect these statistics. As I discussed on page 97, it would be possible simply to make the statistical values properties of the appropriate classes. The disadvantage of this is that, since the counters are dynamic, they will probably need to be fetched from some hardware each time they are read—potentially a time-consuming operation which should be avoided if the user really only wants the static content of a class.

I consequently create a number of statistical classes and associate them with the devices creating the statistics. The class which we will need to subclass is CIM_StatisticalData and it already comes with a number of subclasses. CIM_NetworkPortStatistics seems a useful candidate until its properties are studied: it does not contain general port

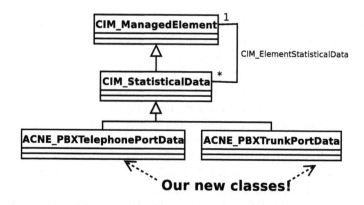

Figure 9.6   The PBX Statistics Classes

statistics, only the port statistics for a port carrying packetised traffic. It appears therefore that we need to create our own statistics classes: see Figure 9.6.

Each statistical class will be associated with its port by the CIM_ElementStatisticalData association (as defined in Figure 6.3 on page 98).

Note that, by inheriting from CIM_StatisticalData, our new class automatically has many useful properties and methods, including the date and time at which the statistics counters were last reset and an extrinsic function to reset them.

Modelling the Events

As you learned in Chapter 8, in order for an operator to receive notification of events, instances of three classes need to be created:

1. A class descended from CIM_Indication which contains details of the event.
2. A class descended from CIM_IndicationFilter to specify precisely which events the operator wishes to receive.
3. A class descended from CIM_IndicationSubscription to tie the filter with the operator's handler.

The only one of these classes which needs to be defined before the PBX is deployed is the descendant of CIM_Indication and so that is the only one which I will consider here. For the other two classes, the ones provided by the DMTF will probably suffice.

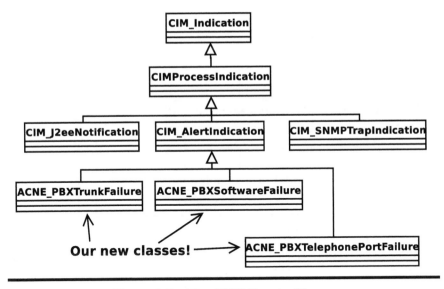

Figure 9.7   The PBX Events Classes

CIM_Indication, as shown in Figure 8.3 on page 155, is subclassed into class, instance, and process indications. The first two deal with events created by the WBEM server itself as classes and instances are created and destroyed. Our interest is in process indications and these are further subclassed as shown in Figure 8.3. Because we are not dealing with SNMP traps, the logical class for us to subclass is CIM_Alert-Indication. All of the properties that we will require, including the time that the event was discovered, the description of the event, the identity of the entity discovering the event, the type and severity of the event and the probable cause of the event, are already available to us in the superclasses and so we simply need to define a class for each event which may occur and in which an operator might have an interest.

I list a few of these in Table 9.1 and illustrate them in Figure 9.7.

Modelling the Services

Finally, we can look at the services we provide: basic telephony, voice mail and diagnostic testing. Anticipating the need for CIM_Service, I included it on Figure 9.4—it is also a subclass of CIM_EnabledLogical-Element. I described the concept of a service on page 92 where I even used the example of voice mail as a prototypical service.

To allow us to concentrate on the service aspects of our PBX, I have redrawn part of the core model in Figure 9.8 Notice that CIM_Diag-

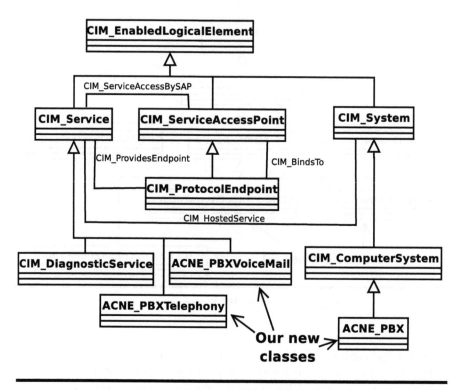

Figure 9.8   Some PBX Service-Related Classes

nosticService is already a subclass of CIM_Service, but that, not unexpectedly, our voice mail and telephony do not yet exist. This gives us the opportunity to introduce two new services: ACNE_PBXVoiceMail and ACNE_PBXTelephony.

We can then use the existing association CIM_HostedService, or a subclass of it, to connect the services to our PBX and we can use CIM_ServiceAccessBySAP or its subclass CIM_ProvidesEndpoint to connect the services with the points through which they may be accessed.

Part of the telephony service requires us to be able to group telephones so that incoming calls can be correctly directed. CIM provides various methods of modelling collections, the grandparent of them all being CIM_Collection (see Figure 9.9). One subclass of this class is CIM_SystemSpecificCollection which is described as follows in its *mof*:

SystemSpecificCollection represents the general concept of a collection which is scoped (or contained) by a System. It represents a Collection that only has meaning in the context of a

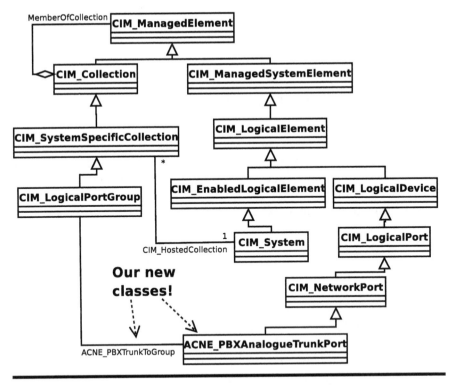

Figure 9.9   The Collection Hierarchy

System, and/or whose elements are restricted by the definition of the System. This is explicitly described by the (required) association, HostedCollection. An example of a SystemSpecific-Collection is a Fibre Channel zone that collects network ports, port groupings and aliases (as required by a customer) in the context of an AdminDomain. The Collection is not a part of the domain, but merely an arbitrary grouping of the devices and other Collections in the domain. In other words, the context of the Collection is restricted to the domain, and its members are also limited by the domain.

This sounds very like our grouping of telephone ports and, in fact, CIM_SystemSpecificCollection actually has a subclass which appears to be even more suitable for our application: CIM_LogicalPortGroup. This class allows us to name the specific collection and, through the CIM_MemberOfCollection association, we can group telephony ports into a particular instance of the class. Having created our group of Telephone Ports, we have to associate a trunk port with the group: for this I have created an association ACNE_PBXTrunkToGroup as shown in Figure 9.9.

Adding Unnecessary Classes

In the previous paragraphs we have added a few of the classes we need to model the PBX. Let us consider briefly adding some classes that we do not need. Take, for example, our power supply. The common models contain a very adequate class with many properties called CIM_Power-Supply which inherits from CIM_LogicalDevice which is precisely where we would introduce our ACNE_PBXPowerSupply. Do we need to define our own class here?

The strict answer is "no"; we can use the one already defined. There might, however, be advantages later in defining a ACNE_PBXPower-Supply class which inherits from CIM_PowerSupply and adds no new features. This would give us flexibility in the future if we did find that our power supply has a property or feature not already modelled.

This is a judgment call that you would need to make on a case by case basis. In this case I believe that I would feel more comfortable with an additional class. For the purposes of this exercise, however, I will not add any unnecessary classes.

Adding Properties

Now that the basic structure of the classes is complete, we can think about the properties we need to add to them. Because this is a much simpler exercise than choosing and positioning classes, and one where errors can be easily corrected, I will not go into as much detail here. I will, however, choose a single class, ACNE_PBXTelephoneModule, and consider the types of property it might have. This will allow me to use this class as as example when writing a provider for it in Chapter 12. The *mof* defining this class is included in that chapter as Figure 12.1 on page 216.

This class was designed to represent the logical view of the telephone interface card (rather than its physical view). As such it might have properties such as the following (note that (*) indicates that a client may modify the property):

```
uint32 Protocol;      // European, N.America, etc. (*)
```

Class	Superclass	Notes
ACNE_PBX	CIM_ComputerSystem	Whole PBX
ACNE_PBXAnalogueTrunkPort	CIM_NetworkPort	Ports
ACNE_PBXAnalogueTelPort	CIM_NetworkPort	
ACNE_PBXProcModule	CIM_LogicalModule	Cards
ACNE_PBXTelephoneModule	CIM_LogicalModule	
ACNE_PBXTrunkModule	CIM_LogicalModule	
ACNE_PBXTelephony	CIM_Service	Services
ACNE_PBXVoiceMail	CIM_Service	
ACNE_PBXTrunkToGroup	CIM_Dependency	Association
ACNE_PBXTelephonePortData	CIM_StatisticalData	Statistics
ACNE_PBXTrunkPortData	CIM_StatisticalData	
ACNE_PBXTrunkFailure	CIM_AlertIndication	Alarms
ACNE_PBXSoftwareFailure	CIM_AlertIndication	
ACNE_PBXTelephonePortFailure	CIM_AlertIndication	

Table 9.1   PBX Classes

These, of course, are in addition to the property it inherits from
CIM_LogicalModule:

```
uint16 ModuleNumber;              // slot number
```

and those which it indirectly inherits from CIM_LogicalDevice:

```
string SystemCreationClassName;  // Key
string SystemName;               // Key
string CreationClassName;        // Key
string DeviceID;                 // Key
uint32 Reset();                  // reset device
uint32 SaveProperties();         // save state
uint32 RestoreProperties();      // restore saved state
```

and those which it inherits from CIM_ManagedSystemElement
(CIM_LogicalElement contains no properties):

```
datetime InstallDate;
string   Name;
uint16   OperationalStatus[];  // degraded, stressed, etc.
string   StatusDescriptions[]; // info about status
```

and, highest of all in the tree, the properties it inherits from CIM_ManagedElement:

```
string Caption;      // description of the object (*)
string Description;  // description of the object (*)
string ElementName;  // user-friendly name (*)
```

Because this is a pedagogical example which I will be using later to demonstrate how to write providers, it would be useful also to have a couple of extrinsic methods. To this end I will therefore assume that the logical device has some form of indicator (an LED on the front-plate perhaps) which can be lit and extinguished on demand. The associated method is:

```
Boolean setIndicator(Boolean newValue);
```

This method accepts a single Boolean parameter and sets the indicator on (true) or off (false), returning the previous value of the indicator.

Summary of our PBX

Over the course of the this chapter, I have addressed some of the technical and commercial points that you will need to consider when building a model. I have sketched the outline of a model for a very simple PBX. In Chapter 12, I look at designing and coding some of the providers for this PBX.

Frequently Asked Questions

FAQ 34 *How long does this modelling work take?*

This really is a "how long is a piece of string?" question. In the team creating the model you will need to combine someone with a detailed knowledge of the product and someone with an exhaustive knowledge of the core and common models. These skills are unlikely to be found in a single person. Then you will have to equip the team with the tools it needs, including tools to move around the *mof* files quickly and easily, trying things out and making changes. Then you will need to give them time to complete the work. For a genuine PBX with many more features than that described here, a four-week period for a team of two or three is probably not unrealistic. Of course, if you intend to take the model to the DMTF to be published as a standard then the amount of work increases significantly.

Chapter 10

Modelling Tips

This section points out a few "gotchas" that exist within the CIM infrastructure. It is not meant to be exhaustive but rather to focus on small points which may be important in building a model: actually places where I have tripped over a point of detail when building a model.

Instances and Classes

Be aware of whether to define a particular entity as a class or as an instance. Superficially you may feel that the distinction is obvious, but consider a 56 kbps dial-up modem from the ABC Manufacturing Company with model number 23/567AB. The DMTF provides a description of a generic dial-up modem in its Device Common Model—see Figure 10.1—so you might consider creating a new class called something like ABC_Type23567AB_Modem and making it a subclass of CIM_POTSModem.* This would probably be wrong! Although there are pathological cases where it would be better to model it as a subset of a CIM_POTSModem, your 23/567AB, 56 kbps modem from ABC Manufacturing is almost certainly an instance of (rather than a subset of) a CIM_POTSModem.

---

* The acronym POTS stands for "plain old telephone system" and a POTS modem is one which works by using in-band tones across a standard telephone line rather than a cable modem or an ADSL modem.

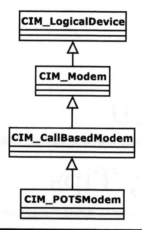

Figure 10.1   CIM_POTSModem and its Superclasses

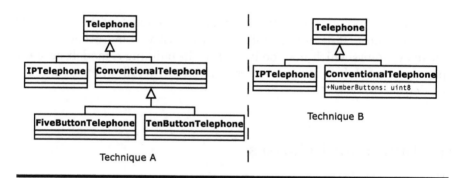

Figure 10.2   Possible Models of Telephones

Subclassing or Defining Types

Another area where a modelling decision needs to be made is whether to add a property to a class to define a subtype or to specify a subclass. Consider, for example, the telephones supported by a small PBX that you are modelling. Assume that there are three types of telephone: a conventional handset with 5 programmable buttons, a conventional handset with 10 programmable buttons and an IP handset (i.e., a handset which, instead of being attached to a conventional telephone socket, is instead connected to a Local Area Network and which sends and receives speech as IP packets). There are probably sufficient differences between the IP and conventional handsets (e.g., the presence of an IP address in addition to the extension number) to define these as two different subclasses of a `Telephone` class; see Figure 10.2.

However, would it be better to model the two types of conventional telephones by creating a further level of hierarchy as shown in Technique A or would it be better simply to add a property to the `ConventionalTelephone` class specifying the number of programmable buttons (Technique B)?

When making this type of decision, consider whether the 5-button telephones and the 10-button telephones will have different associations and different subclasses. If so, then it is probably better to subclass (Technique A); otherwise to use a property. Assuming that there are no differences between the telephones other than the number of buttons, using a property would be better.

Examples of this type of decision can be found throughout the core and common models. The class CIM_OSPFArea, for example, has a property called AreaType which specifies whether the area is normal, a stub or a not-so-stubby-area (NSSA!) (it does not matter what these mean—for the purposes here they are just types of OSPFArea). It would have been possible to subclass CIM_OSPFArea into classes such as CIM_OSPFStubArea, CIM_OSPFNSSAArea and CIM_OSPFNormalArea but the decision was made instead to use a type property. A comment is placed into CIM_OSPFArea's description to annotate this decision: "This class has a 'Type' property, which distinguishes between the different area types. This approach was chosen, because it provides a simpler way to indicate the type of an area, and additional subclassing is not needed at this time."

References

There are a couple of gotchas around references:

Arrays of References are forbidden. The CIM language specification explicitly forbids arrays of references; arrays of any other basic type are legal.

References must be keys. Where references are used (in associations), they must be keys.

If you find either of these restrictions on your use of references constraining, then consider using Object Paths (defined as strings—see page 66) containing the full pathname of an instance instead of a reference. These may require more memory and may be more inefficient to handle but they are normally more flexible.

Underscores in CIM Names

A class name in CIM (e.g., CIM_ComputerSystem) comprises two parts: a schema name and an identifier separated by an underscore. For obvious reasons, the schema name may not contain an underscore but there is confusion about whether the identifier may. That is, is CIM_Computer_System a valid class name?

Appendix F of issue 2.2 of the CIM Specification (document DSP0004) allows identifiers not only to contain underscores but even to begin with one (making CIM__Computer a legal class name). Some *mof* compilers, however, assume that the only underscore in a class name occurs between the schema and the identifier and reject class names with another underscore.

My advice is to keep clear of underscores in class names.

While on the subject of class names, there is also the problem of quotation marks (") in class names. The DMTF specification is also somewhat vague about these and you would do well to avoid them.

Keys

There is one very important property of keys to be found in the CIM Specification:

> If a new subclass is defined from a superclass, and the superclass has key properties (including those inherited from other classes), the new subclass *cannot* define any additional key properties. New key properties in the subclass can be introduced only if all classes in the inheritance chain of the new subclass are keyless.

In the Core Model, for example, CIM_LogicalDevice is defined with four keys:

```
class CIM_LogicalDevice : CIM_EnabledLogicalElement
    {
    [Propagated("CIM_System.CreationClassName"),
        Key, MaxLen (256), Description (
            "The scoping System's CreationClassName.") ]
    string SystemCreationClassName;

    [Propagated("CIM_System.Name"),
        Key, MaxLen (256), Description (
```

```
        "The scoping System's Name.") ]
    string SystemName;

    [Key, MaxLen (256), Description (
        "CreationClassName indicates the name of the "
        "class or the subclass used in the creation "
        "of an instance. When used with the other key "
        "properties of this class, this property allows "
        "all instances of this class and its subclasses "
        "to be uniquely identified.") ]
    string CreationClassName;

    [Key, MaxLen (64), Description (
        "An address or other identifying information "
        "to uniquely name the LogicalDevice.") ]
    string DeviceID;

    // fields deleted
    };
```

This means that no subclass inheriting from CIM_LogicalDevice can have a key other than these four. See also the section on overriding below for more general details on the overriding of keys.

Overrides

In C++, overriding of methods is simple: the superclass may declare a method to be virtual and a subclass may define a method with the same name and signature. The subclass' method then overrides that in the superclass.

In its simplest form, overriding operates much in the same way in CIM and either methods or properties may be overridden:

```
class Superclass
    {
    string Name;
    };

class Subclass : Superclass
    {
    [Override ("Name"), maxLen(15)]
```

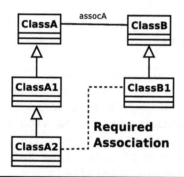

Figure 10.3   Association Overriding

```
string ShortName;
};
```

The property ShortName overrides (i.e., replaces) the property Name when instances of the class Subclass are being addressed. In this definition I have taken the opportunity to restrict the maximum length of the name to 15 characters and this is a very common use of the Override qualifier.

Another common usage is to rename, and possibly restrict, an association. Consider the classes in Figure 10.3. Assume that instances of ClassA (and its subclasses) may have a very general association with instances of ClassB (and its subclasses) as shown: association assocA. Perhaps at this level we can only give very general terms for the two ends of the association such as "antecedent" and "dependent":

```
[Association]
class assocA
    {
    ClassA REF Antecedent;
    ClassB REF Dependent;
    };
```

By the time we reach the subclasses ClassA2 and ClassB1 we want to make use of this association but now we can give much more useful names to the endpoints. We can therefore override the association thus:

```
[Association]
class lowerAssoc : assocA
    {
```

```
[Override ("Antecedent")]
ClassA2 REF LeftHandSide;
[Override ("Dependent"), min(1),max(1)]
ClassB1 REF RightHandSide;
};
```

We have tightened up the names and added the additional restrictions of `max(1)` and `min(1)`.

These two examples illustrate one of the conditions of overriding—you can only constrain the overridden entity more tightly, you cannot loosen it. Thus an unconstrained string as in my first example can have its length constrained to 15 characters or an association without constraints as in my second example can be constrained to `max(1)` but it would not be possible to specify the opposite.

To prevent overriding from constraining a property (or method—anything that I say here about properties is also true of methods) there is a `flavor` which may be attached to a qualifier: `DisableOverride`. For example, the `Association` qualifier is defined as follows:

```
Qualifier Association : boolean = false,
    Scope(association),
    Flavor(DisableOverride);
```

which prevents anything which inherits from a class with the qualifier `Association` from removing that qualifier.

CreationClassName and InstanceID

If you have scanned the *mof* for the core and common models, you will have noticed the property CreationClassName appearing repeatedly. The definition of CIM_LogicalDevice which I have reproduced on page 192 is a particular example.

CreationClassName is something of an embarrassment. It was introduced to keep keys of instances of subclasses unique. Assume, for example, that you have defined classes A, B, and C and that a C *is a* B which *is an* A:

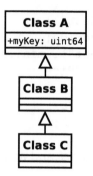

If A defines a key property, myKey, then this is inherited by all instances of classes B and C. If this is the only key, then it is awkward to ensure that instances of classes B and C have unique keys. This problem was worked around by introducing an additional key into classes like A: CreationClassName. An instance of B might then have the keys

```
myKey=27,CreationClassName="B"
```

and an instance of C might have the keys

```
myKey=27,CreationClassName="C"
```

Although both instances have myKey set to 27, they are still unique because of the second key: myKey only needs to be unique between the instances of one class.

Such a key, when used in this manner, also has the effect of providing what in C++ would be called a downcast. If I am given an instance of class A, I can find out by inspecting the keys whether it is also an instance of class B or class C. This can often be useful.

However, the general feeling within the DMTF is that Creation-ClassName is becoming something of a liability—it is proliferating like wild rabbits throughout the CIM schema. There is therefore a move away from CreationClassName towards arbitrary and opaque instance identifiers. An example is given in the class CIM_SettingData which I reproduce in slightly abbreviated fashion here:

```
[Key, Description (
"Within the scope of the instantiating Namespace, "
"InstanceID opaquely and uniquely identifies an "
"instance of this class.  In order to ensure "
"uniqueness within the NameSpace, the value of "
```

```
"InstanceID SHOULD be constructed using the "
"following 'preferred' algorithm:\n"
"<OrgID>:<LocalID>\n"
"Where <OrgID> and <LocalID> are separated by a "
"colon ':', and where <OrgID> MUST include a "
"copyrighted, trademarked or otherwise unique name "
"that is owned by the business entity "
"creating/defining the InstanceID, or is a "
"registered ID that is assigned to the business "
"entity by a recognized global authority (This is "
"similar to the <Schema Name>_<Class Name> structure "
"of Schema class names.)\n"
"If the above 'preferred' algorithm is not used, "
"the defining entity MUST assure that the resultant "
"InstanceID is not re-used across any InstanceIDs "
"produced by this or other providers for this "
"instance's NameSpace.\n") ]
string InstanceID;
```

The InstanceID is therefore a structured but otherwise opaque key. The Acne Manufacturing company could assign InstanceIDs of the form ACNE:0000001, ACNE:0000002, ACNE:0000003, etc., incrementing the number for each created instance.

Actually, as the role of CreationClassName is simply to ensure uniqueness, there is really no need for you to set them to the genuine class name. If you wish, you may set them to anything which makes the instance unique, including the format of the InstanceID described above.

Namespaces

As in C++, a namespace is a "boundary" within which keys must be unique. As illustrated in my section on instance naming, starting on page 66, the namespace actually forms part of the full name of an instance.

Generally, namespace names look like a UNIX directory tree (e.g., root/CIMV2 and root/CIMV2/test) but this is misleading—there is actually neither structure nor hierarchy in these names. The slashes in the names are simply characters and each namespace is independent of all others.

When classes are created and loaded into the WBEM server, their superclasses must exist in the same namespace, which may mean that the CIM schema need to be loaded into several namespaces. An association, however, may relate classes or instances in different namespaces as the namespace can be included in the reference fields.

Boolean Qualifiers

Boolean qualifiers may have defaults of *true* or *false*. If you use a Boolean qualifier without specifying a value then it *does not adopt the value of the default—it becomes "true"!* Make sure that this is what you mean or, better still, always specify a value.

For example, the Abstract and Read qualifiers are defined in the standard core model as follows:

```
Qualifier Abstract : boolean = false,
    Scope(class, association, indication),
    Flavor(Restricted);

Qualifier Read : boolean = true,
    Scope(property);
```

If you define a class as follows:

```
[Abstract, Description("My class")]
class MyClass
    {
    [Read]
    uint8  myValue;
    };
```

then MyClass is Abstract as if you had specified Abstract(true) (even though the default for Abstract is false) and myValue is readable.

Frequently Asked Questions

FAQ 35 *Why can I not change the value of a property if it has the KEY qualifier?*

From a philosophical viewpoint, because it is the combination of values of the key properties that defines a particular instance. If the value of a key property changes then you have created a new instance.

From a practical viewpoint, the WBEM server will use the values of the keys to store instances in such a way that they can be efficiently retrieved. Changing a key of an instance already created would be very difficult to handle.

FAQ 36 *How can I create a unique opaque key—the InstanceID?*

This varies from application to application. The simplest, but often impractical, technique is to have a central authority which issues an incrementing number. It can ensure that the same number is never issued twice.

If this is impractical look for a base number which is guaranteed to be globally unique, a MAC address, for example. Then combine this with something which is locally unique, a slot number or user identity, for example. Together these could form a unique key.

Chapter 11

Writing Providers

This chapter is designed to give you an understanding of the types of provider typically supported by WBEM server implementations. Unfortunately, the DMTF standards define neither what types of provider should be supported by a WBEM server nor what interfaces they should provide although the emergence of the CMPI should resolve this.

Naturally, the types of provider and their interfaces that most implementations support correspond very closely with the intrinsic methods a WBEM client can invoke. For example, because a WBEM client can invoke the `EnumerateInstances()` intrinsic method (see Table 7.1 on page 129), we would expect to see a provider able to handle instances (an "Instance Provider"!) with an interface called something like `EnumerateInstances()`. We find exactly that:

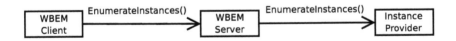

Types of Provider

The following list summarises the types of provider normally supported by a WBEM server; each is described in more detail below.

- **Method Providers** handle calls to extrinsic methods: general-purpose calls to instances of classes. These calls may do anything relevant to the class being invoked.

- **Instance Providers** handle the creation, enumeration (i.e., listing) and deletion of instances of particular classes.
- **Property Providers** handle operator requests for getting and setting properties on an instance.
- **Association (or Associator) Providers** handle the creation, deletion and general handling of associations between classes or instances.
- **Indication (or Indicator) Providers** handle events and alarms raised in the system being managed: filtering them and passing them to programs registered to receive them (this functionality is similar to that of the CORBA event and notification services).
- **Query Providers** handle database-style queries.

Although their functions differ, a single piece of code may implement more than one type of provider: a module may be an instance provider, a property provider and an association provider. Furthermore, many functions can be carried out by different types of provider—modifying a property on an instance might be carried out at different times by a Property Provider, an instance provider or a method provider. The choice of provider type is a design decision.

In general, providers are associated with dynamic entities: counters that are updated by external events, associations that are formed when services are created, etc. Static entities (classes and static instances) generally do not need providers—they can be defined in the *mof* code and loaded once into the WBEM server. The WBEM server will then effectively act as a provider to respond to queries about these classes and instances.

In the C++ world, providers are typically compiled to dynamically linked libraries (.so or .dll) loaded by the WBEM server as required. The WBEM server will then automatically unlink a provider that has not been used for some time (and has not been locked in memory).

Method Providers

A "method" is an extrinsic function defined for a particular class. Such a function may perform any operation on a managed object: a call to a method provider from a WBEM client is effectively a two-way Remote Procedure Call (RPC). The WBEM client specifies the instance on which the procedure is to be invoked, the name of the method and its parameters. The WBEM server invokes the method provider defined for that method for that class and returns the values generated.

If you have CORBA experience, you may interpret this process as a Remote Procedure Call (which it is) and expect the WBEM server to check the type correctness of the parameters and refuse to invoke a method provider if the parameter types are wrong. Generally, WBEM servers do not do this (although the specification is silent on the subject and it is left to the developers of the WBEM server), relying instead on the method provider to do the parameter checking. See FAQ 20 on page 84.

An example might be a method provider for a high-level service, providing a means of shutting the service down:

```
void shutdown(boolean immediately);
```

The same result could be obtained by using a Property Provider and by relying on the side effect of changing a Property's value, as is required in SNMP. For example, you could define a Property called something like `RequiredState` and then allow an operator to set this to a value of `ShuttingDownNow`. A Property Provider could then note this change and, as a side effect of the change, carry out the required function. This is less natural than simply calling a `shutdown(true)` method and much harder to use if complex parameters are involved or if several parameters need to be modified simultaneously (atomically).

The types of parameters which can be passed to a method provider are the normal CIM types: signed and unsigned integers of 8, 16, 32, and 64 bits; strings and characters; Booleans; 4- and 8-octet floating point numbers and dates/times.

A method provider is required to provide a single function: `invoke-Method`.

Instance Providers

An instance provider manages the dynamic instances of a class. In some applications, instances of a class may be created and deleted by an operator. In this case the instance provider may maintain a list of instances (possibly stored in nonvolatile memory). In other applications it may be inappropriate for the operator to be able to create and delete instances (for example, in many applications it makes no sense for an operator to create a new instance of the operating system).

Typically, the instance provider maintains a list of instances either by responding to operator requests or by scanning the real equipment to see what has been fitted. Of course, in the latter case, the list can be "virtual," being recreated from the hardware each time it is required.

For example, when a router is first installed, there is no way for the WBEM server to know *a priori* how many ports of particular types have been installed. An instance provider would communicate directly with the low-level drivers to determine the number of ports. If this information were requested by a WBEM client, then the WBEM server would invoke the instance provider to retrieve it. When instances of a class can be defined statically in advance ("there are always exactly two instances of the power supply: called PS2 and PS2"), they can be configured in a *mof* specification and no instance provider is required, although one could be provided—it would be invoked as the instances defined in the *mof* are compiled and loaded into the repository. Even in this case, an explicit provider might be useful since it could prevent a malicious or faulty WBEM client from creating another instance—the creation request would be routed to the provider which could reject it.

An Instance Provider typically implements the following functions:

- `getInstance()`: Retrieve and return a particular instance
- `enumerateInstances()`: List all available instances
- `enumerateInstanceNames()`: List the names of all instances
- `modifyInstance()`: Modify an instance—similar to the action of a Property Provider
- `createInstance()`: Create an instance (reasonably enough, given the name)
- `deleteInstance()`: Delete an instance (eponymously)

Property Providers

Properties associated with a managed object can be either static or dynamic. A name, for example, might be permanently and statically associated with a device, whereas a field containing a count of the number of packets which have passed through a particular port would be dynamic. This counter changes rapidly and needs to be read from the physical device when requested by an operator. The WBEM server uses a property provider to extract the actual value of the counter only when it receives a request for it.

A Property Provider will typically provide the following methods:

- `getProperty()` or `getPropertyValue()`: Get the value of a property
- `setProperty()` or `setPropertyValue()`: Set the value of a property

As we saw on page 128, the client-side interface does not need to support `getProperty()` and `setProperty()` and some of the implementations of the WBEM servers do not actually support the concept of property providers—relying instead on instance providers to return an instance, including the required properties.

Association Providers

An Association Provider, sometimes called an "Associator Provider," is able to build, destroy, list, and generally manipulate associations between components dynamically.

For example, imagine that an OSPF Service* has been dynamically created on a particular device and that the operator now adds an OSPF Area. This OSPF Area must be linked to the OSPF Protocol Endpoints which it contains. An association is defined in the standard DMTF models for this: `EndPointInArea`.

Because this association is created dynamically, it will be supported by an association provider to keep track of the instances of the association.

This type of provider is an extremely powerful tool for connecting loosely coupled instances as it can answer questions of the type: "With which OSPF Areas is this particular OSPF Protocol Endpoint associated?" or "Through which associations is instance X of class Y associated with instance Q of class R?"

An association provider will typically implement the following functions:

- `associators()`: Return a list of instances associated with (i.e., linked to) a particular instance by a particular association (in case the provider supports more than one type of association). For example, return a list of all objects with which the OSPF Service `myOspfService` is associated.
- `associatorNames()`: Return a list of names of instances associated with a particular instance by a particular association (in case the provider supports more than one type of association). For example, return a list of the names of all objects with which the OSPF Service `myOspfService` is associated.
- `references()`: Return a list of all instances of associations referring to a given instance. For example, what associations connect the OSPF Service `myOspfService` to other objects?

---

* OSPF (Open Shortest Path First) is an IP routing protocol used within a network to distribute routing tables amongst all routers—see Glossary.

- `referenceNames()`: Return a list of the names of all instances of associations referring to a given instance. For example, what associations connect the OSPF Service `myOspfService` to other objects?

Note that, as described on page 139, a WBEM client may invoke an intrinsic method to manipulate associations for either classes or instances. Those invoked on classes ("With which classes is class X associated?") are handled by the WBEM server itself and do not find their way to a provider. The provider described here will only handle requests for instances.

Indication Providers

This provider differs from the others in that it does not simply respond to requests from the WBEM server—it is activated by an external event of some type and initiates communication with the WBEM server. Of all of the provider types this is the one with the least standardisation across WBEM server implementations, although a standardisation process is in progress, driven by the openPegasus, SBLIM, and other teams.

Before reading this section, you may wish to review the whole indication process which I described in Chapter 8, particularly the section on subscriptions starting on page 160. Armed with that information, it is worthwhile considering in a general way the interfaces which an indication provider will require:

- It will require a mechanism to allow it to inform the WBEM server of an indication. This is known as "publishing" in CIM-speak.
- It will require a mechanism that allows it to validate listeners which want to receive its indications ("subscribing" to those indications); because the provider is the authority for notification delivery, there will have to be some way for the WBEM server to ask the provider whether a certain subscription should be allowed. The provider may, in its turn, access some central or distributed authentication and authorisation server.
- Although not essential, for efficiency it will require a mechanism for the WBEM server to inform the provider that no one has subscribed for one or more of the indications it raises. This knowledge would allow the provider to suppress the generation of the indications. Of course, it must also be able to switch the indications back on when a subscription is created.

The last two of these mechanisms are provided by a small number of functions which the indication provider must support:

- `authorizeFilter()`. The WBEM server calls this function when a client creates a subscription. It effectively asks the provider whether this particular client is authorised to receive the indications associated with the filter. How the provider determines this is, of course, outside the specification of the process—it might have a local, compiled-in table, might accept any client or might contact some external authorisation device. The provider might also reject the authorisation on grounds other than the authority of the client—perhaps the filter would be too expensive to use because of its processing time or memory needs.
- `activateFilter()`. This function is also called by the WBEM server when a client creates a subscription for one or more of the indications for which the provider is responsible. If this is the first subscription for a particular indication then the provider can take this as a request to start generating notifications of the indication.
- `deActivateFilter()`. This is, of the course, the opposite of `activateFilter`. It tells the provider that a subscription has been deleted and, if that was the last subscription for a particular indication, allows the provider to stop generating notifications of the indication.
- `mustPoll()`. This is a dangerous function call. It is designed to allow particularly simple indication providers to be written—if the provider responds to this call with *true*, then, instead of it having to generate indications asynchronously, it will be called by the WBEM server at regular intervals (through the instance provider interfaces) to see whether an alarm has occurred. It does not need much imagination to see how this could very quickly become inefficient if misused.

The first mechanism on my list above, that of actually informing the WBEM server that an indication has been raised, is covered by allowing the indication provider to create an instance of the appropriate indication class—effectively to act as a WBEM client and create an instance. At present this is achieved in different ways in different WBEM server implementations.

Query Providers

This type of provider handles the ExecQuery() intrinsic method which can be invoked by a WBEM client (see page 139). It typically provides only one interface: `ExecQuery()`.

Provider/WBEM Server Interfacing

As illustrated in Figure 4.4 on page 39, the interface between the WBEM server and its providers is implemented as a Provider Protocol Adaptor. In the early days of WBEM/CIM these adaptors were not seen as plug-in components and the interface at this point was defined by the particular WBEM server implementation. This led to fragmentation as WBEM servers implemented in Java could not interface with providers written in C++ and *vice versa*. This was particularly a problem with very small devices where it was necessary to write providers in C, memory being inadequate for either C++ or Java.

Gradually a specification, the Native Provider Interface (NPI), emerged to address the issue of interfacing with providers written in C, particularly from Java WBEM servers.

NPI is currently being superseded by the Common Manageability Programming Interface (CMPI). The CMPI specification can be found at `http://www.wbemsource.org`.

Both the openPegasus and OpenWBEM servers currently provide C++ interfaces which are different from the CMPI and each other; but both also now have CMPI adapters.

The CMPI specification includes the following requirements which it is intended to satisfy (my paraphrasing from the CMPI specification):

- Reduce the complexity of writing Management Instrumentation (i.e., providers).
- Allow providers to be written without having to include libraries specific to the WBEM server being used—a C header file being all that is needed.
- Provide support for providers remote from the WBEM server.
- Place no requirement on the providers to maintain state between calls so that providers implemented as simple scripts can be supported.
- Allow multithreading of providers by making the interface thread-safe.
- Support any number of providers in the same library module.

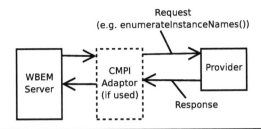

Figure 11.1   WBEM Server/Provider Interface

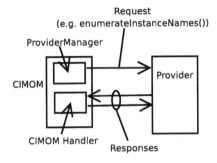

Figure 11.2   openPegasus WBEM Server/Provider Interface

The WBEM server/provider interface works generically on the principle, illustrated in Figure 11.1, of a client/server relationship, with the WBEM server being the client *(sic)* requesting services of the provider, acting as the server.

In its non-CMPI implementation, the openPegasus extends this client-server model and makes the WBEM server and the provider more independent by supporting the interfaces shown in Figure 11.2. The WBEM server loads the provider when it is first required and invokes the provider by giving it a call-back address of a handler. The provider does its work and makes calls to the handler to answer the request. This interaction, sophisticated enough to allow the provider to return its information incrementally, is illustrated in the example in Chapter 12. In principle, incremental delivery from the Provider to the WBEM server allows the WBEM server also to deliver the response to the WBEM client incrementally, using the HTTP "chunked transfer" feature. Unfortunately, because the CMPI specification does not support incremental delivery from the provider to the WBEM server, this rather attractive openPegasus feature is not available in the CMPI implementation.

Chapter 12

Implementing Providers: Example

Implementing Providers: General Steps

This section contains a brief introduction to the steps which you need to take when writing a provider. As the actual details vary from WBEM server to WBEM server, an example is given in the rest of this chapter using one particular WBEM server.

When writing a provider you will need to:

1. Write the *mof* code for the class for which you need a provider. Load this class definition into the WBEM server.
2. Write some code to perform the necessary provider functions. Compile this code into a library and place this into a directory from which the WBEM server can extract it as needed.
3. Inform the WBEM server of the presence of the library and tie it to the class which it supports and the functions which it performs.

These steps are illustrated in the example below.

The Example

This chapter provides an artificial example to show how providers are written using one particular WBEM server, openPegasus. The code in this chapter was written to work with version 2.2 of the openPegasus software but it is general enough that it should work correctly with later releases. Because of the lack of standards surrounding the provider interface, it would not work without modification with other WBEM servers but the basic structure is common to all WBEM server implementations.

Installing openPegasus is remarkably easy and, if you want to follow along with this example on your own computer, then I suggest that you install it—see Appendix F for a detailed description of its retrieval and installation process.

This is not a book about programming in C++ and I have assumed that you will not need me to write out header files and the complete programs for every example. Instead I give the code snippets which handle the interaction with the WBEM server and leave the rest to your imagination and experience.

Some of the code in this chapter has been adapted from the examples released with openPegasus and may therefore be subject to the licencing terms described in Appendix H.

As an example I intend to take one of the classes which I developed in Chapter 9. ACNE_PBXTelephoneModule's origin and *raison d'être* are described on page 177 and its properties and extrinsic methods on page 184. To save you from having to (re)read that chapter I briefly recap here the underlying logical device (telephone interface card) which the ACNE_PBXTelephoneModule represents. You should, however, refer to Figure 9.2 on page 169:

- Up to eight telephone interface cards may be present in a PBX.
- Each interface card supports connections to eight telephones.
- Interface cards may be inserted into slots 3 to 10 of the PBX shelf.

An operator may enquire about the number of telephone interfaces present (but may not create an instance of the class—instances are created only by an appropriate card being inserted into the PBX), may invoke the reset(), saveProperties(), restoreProperties() and setIndicator() extrinsic functions and read all properties.

A Brief Introduction to openPegasus

In order to understand the rest of this chapter, you will need to know a little about the way in which the openPegasus WBEM server handles providers. openPegasus is code made available from **www. openpegasus.org** under the conditions of the licence described in Appendix H. All code is written in C++ and it is anticipated, at least until the CMPI interface is available, that providers will also be written in C++.

Providers are loaded from shared library files (.so) when first invoked and, in version 2.2 of the openPegasus code, run in the WBEM server's address space. This is unfortunate because a faulty provider can bring the server down and it is planned that this will be resolved in a later openPegasus release. A particular provider may be flagged as "unloadable" and will be unloaded by the WBEM server following a preset period of inactivity: this can be particularly useful for providers which are invoked once at system initialisation and never again but might be inappropriate for providers which contain a great deal of state. Such state would need to be stored before the provider was unloaded and reloaded when it was again invoked.

The provider interface used for this example follows the interchanges shown in Figure 11.2 on page 209: the openPegasus WBEM server passes an address of a handler (not to be confused with an indication handler—this is just the normal C++ use of the term handler) to the provider with the request; the provider is expected to call **handler.processing()** before starting to return the response, then **handler.deliver()** any number of times with the (parts of the) response and finally **handler.complete()** to indicate that the response is complete. At this point the handler address becomes stale and must not be used.

Note that, at least in openPegasus version 2.2, the handler is not guaranteed to be thread-safe, so the whole provider must be implemented in a single thread. This is not an issue for the example in this chapter as it is extremely simple but it is something you may wish to consider for larger systems.

The openPegasus program suite contains a program for manipulating running providers, **cimprovider**, which allows you to disable (and thereby unload) specific providers. This is useful during debugging when it may be necessary to force the loading of a new version of a provider. The most useful commands are:

cimprovider -l -s: List all providers known to the WBEM server with their current status.

cimprovider -d -m providerName: Disable the named provider.

cimprovider -e -m providerName: Enable the named provider.

cimprovider -r -m providerName: Remove the named provider.

As a point of detail, it is perhaps worth noting that the default repository for openPegasus is a set of directories and files on disc. Although this makes debugging easy and allows the repository to persist while the WBEM server is unloaded and reloaded, it can sometimes be a nuisance. If, during debugging, you have made a mess of your *mof* code and want to clean everything out, then its persistence can be annoying. To clean everything out of the repository, it is only necessary to go into $PEGASUS_HOME/Schemas and type make clean.

Step 1: Write and Compile the *mof*

The first step of those given on page 211 is to write and compile the *mof* code for the managed object. I show the *mof* for the ACNE_PBXTelephoneModule class in Figure 12.1 and have typed it into the file moduleClass.mof.

Before we can invoke the *mof* compiler we must select a namespace for our classes. I have assumed the namespace root/acnePbx. If we try to compile our classes into this namespace then we will get an error, since the CIM_LogicalModule class is unknown in that namespace. We must therefore start by compiling the CIM schema into root/acnePbx. Each set of CIM schema as released by the DMTF has one top-level file, which consists of a series of #pragma include statements which collectively cause the whole of the CIM schema to be loaded. It is therefore only necessary to compile (load) this one file into the WBEM server. This file is normally known as CIM_SchemaXX.mof where XX is the version number.

Two versions of the openPegasus *mof* compiler are available to compile our two *mof* files and load their contents into the repository:

```
cimmof -Rlocalhost:5988 -uc -aEV -aV -nroot/acnePbx
                        CIM_Schema28.mof
cimmof -Rlocalhost:5988 -uc -aEV -aV -nroot/acnePbx
                        moduleClass.mof
```

or

```
cimmofl -R<repository name> -nroot/acnePbx CIM_Schema28.mof
cimmofl -R<repository name> -nroot/acnePbx moduleClass.mof
```

Both cimmof and cimmofl have a number of useful flags, which are listed by running the programs with the -h flag. These include:

-Rserver_port: Tells `cimmof` where to find the WBEM server. An example might be "-Rlocalhost:5988" to use a WBEM server on the local computer using an HTTP (as opposed to HTTPS) connection.

-uc: Allows a class which is already known by the WBEM server to be replaced by a new version (but only if its version number is higher than the old one).

-aEV: Allows experimental classes (i.e., classes which have not been formally accepted into the standard—see page 88) to be compiled.

-aV: Allows both major and backward revision changes to be compiled. Normally you would expect version numbers of a class to increase monotonically: 2.7.5 going to 2.8.0 and so on. In this case, a decrease in version number would be an error. By default, therefore, `cimmof` and `cimmofl` will not allow a class definition to be replaced by one of the same name but with lower version number. Similarly, a major version change would normally require a software upgrade and this would also represent an error. The -aV flag tells `cimmof` and `cimmofl` to accept the new class definition even if these strange version number conditions occur.

-nnamespace: Specifies the namespace into which the classes and instances are to be loaded.

Both `cimmof` and `cimmofl` check the syntax of `moduleClass.mof`, but differ in what they do with the resulting complied output. `cimmof` is a full WBEM client and passes the new class to the WBEM server through the standard CIM-XML client interface, whereas `cimmofl` (i.e., *cimmof Local*) writes the information directly into the repository. To this extent, `cimmofl` is dangerous, as any associated providers will not be invoked. If you are confident that you know what you are doing, then `cimmofl` is useful because it can be used even if no WBEM server is running, but generally it should be avoided; it is really only meant as a debugging tool.

It is instructive to view the XML that the `cimmof` compiler generates for the *mof* code shown in Figure 12.1 on the following page. The XML is shown in Figure 12.2 on page 217—it complies with the "Specification for the Representation of CIM in XML" document produced by the DMTF (document DSP0201). If nothing else, comparing the two representations makes one realise the inefficiency of the XML notation.

While we are writing *mof* for this example, it is perhaps worth creating an instance of the ACNE_PBX class. Although it is not strictly

```
[Version ("4.5.3"), Description (
    "ACNE_PBXTelephoneModule represents the logical function "
    "of the 8 port telephone interface card fitted in the "
    "model 34/AGB/76 PBX.") ]
class ACNE_PBXTelephoneModule : CIM_LogicalModule
    {
    [Write,
    Description ("Telephone Protocol"),
    ValueMap { "0", "1", "2", "3" },
    Values {"UK", "European", "NorthAmerican", "Japanese" } ]
    uint32 Protocol;

    Boolean setIndicator(Boolean newValue);
    };
```

Figure 12.1 *mof* **Code for the Telephone Module**

necessary for the provider example to work, it does give us another instance to enquire about. The code for this is given in Figure 12.3.

Finally, for the purposes of our example, we need to consider the manner in which the keys of the instances of ACNE_PBXTelephoneModule will be allocated. As we have no real telephone hardware, our provider will have to pretend, using instances preset in a table. The keys of ACNE_PBXTelephoneModule are, of course, inherited from CIM_LogicalDevice and, for the purposes of this example, I have assumed that there are three instances of ACNE_PBXTelephoneModule and that their keys are as follows:

SystemCreationClassName. Set to ACNE_PBX for all three instances.

SystemName. Set to XYZCopmanyPBX1 for all three instances.

CreationClassName. Set to ACNE_PBXTelephoneModule for all three instances.

DeviceID. Set to 1, 3, and 6 for instances 1, 2, and 3 respectively.

Note that three of the four keys are identical in all cases so the instances are really only being differentiated by DeviceID.

```
<?xml version="1.0"?>
<!-- Open Group Pegasus CIM Compiler V 2.2 Built Jul 11 2003 -->
<CIM CIMVERSION="2.0" DTDVERSION="2.0">
<DECLARATION>
 <DECLGROUP>
  <VALUE.OBJECT>
   <CLASS NAME="ACNE_PBXTelephoneModule"
                         SUPERCLASS="CIM_LogicalModule">
    <QUALIFIER NAME="Version" TYPE="string" OVERRIDABLE="false"
                         TOSUBCLASS="false">
     <VALUE> 4.5.3 </VALUE>
    </QUALIFIER>
    <QUALIFIER NAME="Description" TYPE="string" OVERRIDABLE="false"
                         TOSUBCLASS="false">
     <VALUE> ACNE_PBXTelephoneModule represents the logical
                function of the 8 port telephone interface card
                fitted in the model 34/AGB/76 PBX. </VALUE>
    </QUALIFIER>
    <PROPERTY NAME="Protocol" CLASSORIGIN="ACNE_PBXTelephoneModule"
                         TYPE="uint32">
     <QUALIFIER NAME="Write" TYPE="boolean" OVERRIDABLE="false"
                         TOSUBCLASS="false">
      <VALUE> TRUE </VALUE>
     </QUALIFIER>
     <QUALIFIER NAME="Description" TYPE="string"
                         OVERRIDABLE="false" TOSUBCLASS="false">
      <VALUE> Telephone Protocol </VALUE>
     </QUALIFIER>
     <QUALIFIER NAME="ValueMap" TYPE="string" OVERRIDABLE="false"
                         TOSUBCLASS="false">
      <VALUE.ARRAY>
       <VALUE> 0 </VALUE>
       <VALUE> 1 </VALUE>
       <VALUE> 2 </VALUE>
       <VALUE> 3 </VALUE>
      </VALUE.ARRAY>
     </QUALIFIER>
     <QUALIFIER NAME="Values" TYPE="string" OVERRIDABLE="false"
                         TOSUBCLASS="false">
      <VALUE.ARRAY>
       <VALUE> UK </VALUE>
       <VALUE> European </VALUE>
       <VALUE> NorthAmerican </VALUE>
       <VALUE> Japanese </VALUE>
      </VALUE.ARRAY>
     </QUALIFIER>
    </PROPERTY>
    <METHOD NAME="setIndicator" TYPE="boolean">
     <PARAMETER NAME="newValue" TYPE="boolean"> </PARAMETER>
    </METHOD>
   </CLASS>
  </VALUE.OBJECT>
 </DECLGROUP>
</DECLARATION>
</CIM>
```

Figure 12.2   XML Generated from the *mof* in Figure 12.1

Step 2: Write the Provider Code

The second step is to write the code for the providers. We must first decide which providers are required—we will clearly need at least an instance provider and a method provider. It would be possible to imagine other providers but, once you have written one type, the others are very simple and so I only give these two examples in detail. The open-Pegasus release comes with examples of all types of providers and these should be the basis of your code. We will write two C++ programs:

- PbxMain.cpp which simply acts as the main program for the providers and creates instances of the Instance Provider and Method Provider classes when requested by the WBEM server. This program is trivial; the code is given in Figure 12.4 and I do not discuss it further below.
- PbxTelephoneModule.cpp (and the associated PbxTelephone-Module.h header file) which provides the functions necessary for the Instance and Method Providers in two classes: PbxInstanceProvider and PbxMethodProvider.

Instance Provider

The instance provider will need to support a number of interfaces:

- getInstance(): Pretend to access the logical device on the PBX and create and return an instance of the ACNE_PBXTelephoneInterface class from the information it discovers.
- enumerateInstances(): Pretend to access the hardware and return all available instances.
- enumerateInstanceNames(): Do the same as enumerateInstances() but return only a list of the names of the instances.
- modifyInstance(): Determine whether or not the client is allowed to modify the properties (is it writable?) and, if so, implement the change.
- createInstance(): Always return an exception since instances of this class may not be created by the client, only by having a card plugged into a slot where they are detected by the PBX software.
- deleteInstance(): Also always return an exception—instances are deleted only by the appropriate card being removed from the system.

The easiest methods to code are those which do nothing! As we have seen, `createInstance()` and `deleteInstance()` should actually only return an error as the operator is not allowed to create or delete instances of the PbxTelephoneModule class—instances being created and deleted by the physical plugging in of cards.

The code for `createInstance()` is given in Figure 12.5 and, since the code for `deleteInstance()` is identical, I have not given it.

You may query the choice of exception: CIMNotSupportedException does not really describe the problem—the `createInstance()` method is forbidden rather than unsupported. Unfortunately the CIM specification (DSP0200) includes only a limited number of exceptions that may be thrown* and has no facility for you to add more. It therefore often arises that you will have to make the best of a bad job with the choice of exception and redeem yourself with the accompanying message. The poverty of exceptions has been recognised by the DMTF and a new class, CIM_Error, was introduced in version 2.8 of the CIM schema. The idea is that, in the future, an instance of this class, which has properties such as ErrorType, OwningEntity, PerceivedSeverity, ProbableCause, will accompany the exception.

The methods that actually do something are not much more difficult and I will only consider `enumerateInstanceNames()` and `getInstance()` in this book as the other routines are very similar and are largely built from components contained in these two.

Consider `enumerateInstanceNames()`. Because we have no real hardware, I have invented three instances of the modules by defining them as constants—in real life, of course, the code would go to the hardware to find out what was really installed. The three instances are defined in `PbxTelephoneModule.h` as follows and you must accept the lie that they have been created in this form by some program which scanned the hardware:

```
const unsigned int numberInstances = 3;

const char * const instances[numberInstances] =
{ "ACNE_PBXTelephoneModule."
        "SystemCreationClassName=\"ACNE_PBX\","
        "SystemName=\"XYZCompanyPBX1\","
        "CreationClassName=\"ACNE_PBXTelephoneModule\","
        "DeviceID=\"1\"",

  "ACNE_PBXTelephoneModule."
```

---

* In C++, exceptions are "thrown" and "caught" rather than being "raised."

```
          "SystemCreationClassName=\"ACNE_PBX\","
          "SystemName=\"XYZCompanyPBX1\","
          "CreationClassName=\"ACNE_PBXTelephoneModule\","
          "DeviceID=\"3\"",

  "ACNE_PBXTelephoneModule."
          "SystemCreationClassName=\"ACNE_PBX\","
          "SystemName=\"XYZCompanyPBX1\","
          "CreationClassName=\"ACNE_PBXTelephoneModule\","
          "DeviceID=\"6\"" };
```

Note that these keys align with those described earlier and that I have made use of the C++ trick of breaking a string across two lines by using trailing and leading quotation marks. Once you have worked out the quotation marks, you will find that each of the three entries is an Object Path (see page 67) in the form

```
<classname>.<key>=<value>,<key>=<value> ....
```

The code for enumerateInstanceNames() is now easy to write; see Figure 12.6.

The pattern for writing a provider for an intrinsic method is surely now becoming clear to you. I will finish with one further example: getInstance(). Again, as we have no real hardware, I have written a helper function to pretend to build the necessary instance. This is illustrated in Figure 12.7 where the addProperty() method of the CIMInstance class is repeatedly invoked to create an instance. I have not taken the trouble to set values for every property of ACNE_PBXTelephoneModule but I have set most of them.

There is one further helper function which getInstance() uses: checkKeys, which I have listed in Figure 12.8. It simply checks whether an Object Path is a valid key for our ACNE_PBXTelephoneModule class. This routine, like the buildInstance() routine above, is not strictly provider code, but I have included it for completeness and to expose you to a few more openPegasus functions, particularly the support for arrays.

The actual code of getInstance() uses these two routines to return an instance: I give the code in Figure 12.9. The actual work is done in the three calls to handler.

Method Provider

There are four extrinsic methods which need to be provided, three of which our class has inherited from CIM_LogicalDevice:

uint32 Reset() which will cause us to reset the card by pretending to write into a particular register on a particular integrated circuit on the card. The return value is zero to indicate that the reset was completed successfully.

uint32 SaveProperties() and RestoreProperties() which should throw an exception because I have assumed that there is no state to save and restore.

Boolean setIndicator(Boolean newValue) which will cause us to pretend to write into a particular register on the card and return the old value of the indicator.

Apart from constructors, destructors and other start-up and close-down routines, a method provider is expected to provide a single function: invokeMethod() to which all incoming extrinsic method invocations are forwarded. Our code for this is contained in Figure 12.10.

As you can see, this code simply checks the name of the extrinsic function (setIndicator or reset in this case) and invokes a private function to do the work. Notice again the call to handler.processing() followed by a number (here one) call to handler.deliver() followed by a final call to handler.complete(). This is the same as for all openPegasus providers.

Although strictly you have now seen everything about the method provider interface, I have included the private functions in Figure 12.11 and Figure 12.12 as this gives me an opportunity to showcase a few more openPegasus calls: those to manipulate CIMObjectPaths and CIMKeyBindings. Again, remember that there is no real PBX hardware so they only pretend to do something.

Step 3: Tie the Provider Code to the PBX Class

Having written and compiled the provider code, we have to tell the WBEM server that it exists and when to invoke it. This is done by creating instances of three classes as illustrated informally as follows:

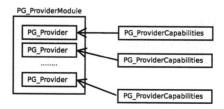

1. **PG_ProviderModule.** This class, which *is a* CIM_Log-
   icalElement, defines a so-called ProviderModule—a group of
   providers collected together into one dynamically linked library
   (.so or .dll file). Assuming that the instance and method
   providers are included in the same library file, the instance required for the PBX example is shown in Figure 12.13.
   Most of the properties in that example are self-explanatory,
   but note the following:
   InterfaceVersion defines the actual version of the openPegasus Application Program Interface (API); this is not a version number which you can choose freely.
   Location defines the actual library.

   $$Location = "PBXLogicalCode"$$

   means that the dynamically linked library is `libPBXLog-
   icalCode.so` in whichever directory is contained in the en-
   vironment variable LD_LIBRARY_PATH.
   InterfaceType defines the interface which the provider will
   use. Currently, openPegasus only supports its own C++
   interface (known as `C++Default`) but in future will offer
   CMPI and other interfaces.
2. **PG_Provider.** An instance of PG_Provider (which also *is a*
   CIM_LogicalElement) defines each specific provider within a
   ProviderModule. Figure 12.14 shows the *mof* for the instance
   and method providers in the PBX example. Note that the Provi-
   derModuleName refers back to the name of the ProviderModule
   defined in Figure 12.13.
3. **PG_ProviderCapabilities.** An instance of this class (which *is
   a* CIM_ManagedElement) defines the precise capabilities of the
   provider—what type of provider is it and what properties or
   methods does it support? The instances for the PBX example
   are shown in Figure 12.15 and Figure 12.16.
   Again, most of the properties are self-explanatory but note
   that a provider may, for example, be both an instance provider
   and a property provider. The types of provider supported by the

code are listed in the ProviderType property—an enumerated value as described in the comments in Figure 12.15.

Once these instances have been defined, they need to be loaded into the repository through the WBEM server. This is achieved by putting them into a file named, say, pbxRegistration.mof, and invoking the compiler (note that the PG_InterOp namespace is used for these "linkage" instances):

```
cimmof -nroot/PG_InterOp pbxRegistration.mof
```

As you can imagine, it is very easy to get the linkages wrong between the PG_ProviderModule, PG_Provider, PG_ProviderCapabilities, and the actual C++ libraries. As the linkages are passed as constant strings, there is no hint of a problem during the compilation process and the only indication is when the provider does not get invoked as expected—a misspelling in the *mof* can take hours to track down. It would be useful to have a program which cross-checked these names but I have not found one. In the meantime use plenty of **printf** or **cout <<** statements to ensure that the providers are being invoked.

Invoking the Providers

Missing Provider

Before actually linking the provider code into the shared library, it is instructive to try to invoke a provider when it is not in the library. I wrote a simple XML file to invoke the intrinsic method EnumerateInstances() on the ACNE_PBXTelephoneModule class while the library file was not present in the library directory. I then used the **wbemexec** utility to send the XML to the WBEM server. The **wbemexec** program is very simple and is provided with the openPegasus release—it connects to the WBEM server as a WBEM client, reads the specified XML file and passes the XML to the WBEM server. It then prints out the XML response which it receives. When I tried to invoke the EnumerateInstances() intrinsic function, not unreasonably I received the message:

```
<?xml version="1.0" encoding="utf-8"?>
<CIM CIMVERSION="2.0" DTDVERSION="2.0">
 <MESSAGE ID="25000" PROTOCOLVERSION="1.0">
```

```
<SIMPLERSP>
 <IMETHODRESPONSE NAME="EnumerateInstances">
  <ERROR CODE="7"
   DESCRIPTION="CIM_ERR_NOT_SUPPORTED: The requested "
              "operation is not supported: "
              "&quot;Invalid provider interface.&quot;"/>
 </IMETHODRESPONSE>
 </SIMPLERSP>
 </MESSAGE>
</CIM>
```

The line breaking in the description of the error is mine to ensure that it was readable in this book. This is a sensible error message given the condition: it simply says that it could not find the provider.

Providers Throwing Exceptions

Having determined that, quite reasonably, the system does not work unless we write providers, let us now compile the providers that we wrote above and add these to the library.

We started our coding with the methods that did not do anything other than throw an exception. Let us start there again and try to create an instance of the ACNE_PBXTelephoneModule class. We can see from the code for **createInstance()** included as Figure 12.5 that this should fail and an exception be thrown.

I give the XML which I sent to the WBEM server in this forlorn attempt to create an instance of ACNE_PBXTelephoneModule in Figure 12.17. Because the input XML is very similar for different commands, I save a few trees by only giving this one example.

As expected, the following XML snippet is returned, surrounded by the normal CIM-XML pre- and postamble (again, the line is broken to make it readable):

```
<IMETHODRESPONSE NAME="CreateInstance">
  <ERROR CODE="7"
   DESCRIPTION="CIM_ERR_NOT_SUPPORTED: The requested "
              "operation is not supported: "
              "&quot;Modules may not be created "
              "manually&quot;"/>
</IMETHODRESPONSE>
```

EnumerateInstanceNames()

We should now try to invoke a method which actually does something other than throw an exception. `enumerateInstanceNames()` is a good example. The XML returned from invoking our `enumerate-InstanceNames()` method is as follows (again excluding the XML pre- and postamble and giving only one of the three instances):

```
<IMETHODRESPONSE NAME="EnumerateInstanceNames">
 <IRETURNVALUE>
  <INSTANCENAME CLASSNAME="ACNE_PBXTelephoneModule">
   <KEYBINDING NAME="CreationClassName">
    <KEYVALUE VALUETYPE="string">
    ACNE_PBXTelephoneModule
    </KEYVALUE>
   </KEYBINDING>
   <KEYBINDING NAME="DeviceID">
    <KEYVALUE VALUETYPE="string"> 1 </KEYVALUE>
   </KEYBINDING>
   <KEYBINDING NAME="SystemCreationClassName">
    <KEYVALUE VALUETYPE="string"> ACNE\_PBX </KEYVALUE>
   </KEYBINDING>
   <KEYBINDING NAME="SystemName">
    <KEYVALUE VALUETYPE="string"> XYZCompanyPBX1 </KEYVALUE>
   </KEYBINDING>
  </INSTANCENAME>
  ......(for other instances).....
 </IRETURNVALUE>
</IMTHODRESPONSE>
```

This is effectively just the Object Path encoded in XML.

GetInstance()

Given the rather artificial code for `getInstance()`—see Figures 12.7 to 12.9, it is easy to predict the XML that a call to GetInstance will return to the WBEM client (again shorn of its XML wrappers):

```
<IMETHODRESPONSE NAME="GetInstance">
 <IRETURNVALUE>
  <INSTANCE CLASSNAME="ACNE_PBXTelephoneModule">
   <PROPERTY NAME="SystemCreationClassName" TYPE="string">
    <VALUE>ACNE_PBX</VALUE>
```

```
    </PROPERTY>
    <PROPERTY NAME="SystemName" TYPE="string">
     <VALUE>XYZCompanyPBX1</VALUE>
    </PROPERTY>
    <PROPERTY NAME="CreationClassName" TYPE="string">
     <VALUE>ACNE_PBXTelephoneModule</VALUE>
    </PROPERTY>
    <PROPERTY NAME="DeviceID" TYPE="string">
     <VALUE>3</VALUE>
    </PROPERTY>
    <PROPERTY NAME="Protocol" TYPE="sint32">
     <VALUE>1</VALUE>
    </PROPERTY>
    <PROPERTY NAME="ModuleNumber" TYPE="sint32">
     <VALUE>4</VALUE>
    </PROPERTY>
    ....and so on for the remaining properties
   </INSTANCE>
  </IRETURNVALUE>
</IMETHODRESPONSE>
```

This is just a list of the property values which we hard-coded into the helper functions for the provider.

Extrinsic Method Invocation

As you can see from Figure 12.11 and Figure 12.12, I actually wrote the code for two extrinsic functions: `reset()` and `setIndicator()`. Of these, `setIndicator()` is by far the more exciting[†] as it can be provoked to return exceptions. If, for example, it is invoked with two parameters instead of the one expected, then the following is returned to the WBEM client (again, I have broken the string for readability):

```
<METHODRESPONSE NAME="setIndicator">
 <ERROR CODE="4" DESCRIPTION="CIM_ERR_INVALID_PARAMETER: "
        "One or more parameter values passed to the "
        "method were invalid: "
        "&quot;setIndicator needs 1 param&quot;"/>
</METHODRESPONSE>
```

[†] This depends, of course, on your standards of excitement.

This incorporates the error string which I programmed into the code.

If this method is invoked correctly with a single parameter set to TRUE then it returns FALSE:

```
<METHODRESPONSE NAME="setIndicator">
 <RETURNVALUE PARAMTYPE="boolean">
  <VALUE> FALSE </VALUE>
 </RETURNVALUE>
</METHODRESPONSE>
```

Implementing Indication Providers

I have left the indication provider until last because it is different from the other providers—it does not wait passively for the WBEM server to invoke it; instead it waits patiently for an external event to occur and then invokes the WBEM server.

openPegasus does not yet conform to the complete indication provider interface described on page 207; authorizeFilter(), (de)activateFilter(), mustPoll() are not supported. As these will appear in openPegasus in the near-future and invalidate everything in the section, I have kept the description of the current interface short.

The indication provider code is very similar to that of a "normal" provider. The first routine of importance is enableIndications() which the openPegasus WBEM server invokes to inform the provider that it should start sending indications (someone has subscribed for them). The parameters to this function include a call-back object in the WBEM server to be used when an event occurs. Normally, it is enough just to store these locally:

```
void MyIndicationProvider::enableIndications (
    IndicationResponseHandler & handler)
    {
    enabled = true;
    myHandler = &handler;
    return;
    }
```

The provider then awaits an event which should cause an indication to be raised and simply creates it:

```
CIMInstance indicationInstance(CIMName("MyIndication"));
```

```
// create an instance of MyIndication class
CIMObjectPath path ;
path.setNameSpace("root/SampleProvider");
path.setClassName("MyIndication");

indicationInstance.setPath(path);

...create and add the properties....
...e.g., indicationInstance.addProperty

CIMDateTime currentDateTime =
                    CIMDateTime::getCurrentDateTime();
indicationInstance.addProperty(CIMProperty("IndicationTime",
                                        currentDateTime));

Array<String> correlatedIndications;
indicationInstance.addProperty(CIMProperty(
                                "CorrelatedIndications",
                                correlatedIndications));

// deliver the instance we have created
myHandler->deliver(indicationInstance);
```

This is sufficient (perhaps with a few more properties set to give more information about the event) for the indication to reach the handler, listener and consumer.

Summary of openPegasus Providers

openPegasus currently expects providers to be written in C++. It provides a method of tying providers for classes to C++ routines in a library. When the WBEM server needs a provider it loads the code dynamically.

Writing one's first provider is quite error prone (e.g., linkage to the *mof*), but openPegasus provides many useful routines to manipulate lists, object paths, strings, etc. It also has many useful example providers. Writing one's second provider is easy: lots of cut and paste.

Frequently Asked Questions

FAQ 37 *What is the minimum that a correctly written openPegasus provider needs to do?*

When invoked, it needs either (1) to call `handler.processing()` and `handler.complete()` or (2) to throw an exception (in which case neither `handler.processing()` nor `handler.complete()` are necessary). There is no requirement for it to call `handler.deliver()` unless it has something to deliver back to the client.

FAQ 38 *Your examples make use of openPegasus methods such as constructors for CIMObjectPath objects. Where can I find a list of the available classes and methods?*

There is no really good documentation for the openPegasus project. This has been accepted by the developers, and as industry acceptance of openPegasus grows, it will change. In the interim there are two main sources to search for information: the class and method list generated by Doxygen is available from the openPegasus Web site and the example and test programs which come with the openPegasus source. There are numerous test programs used after each build to check the integrity of the resulting WBEM server and these contain many examples of both valid and invalid method calls.

The header (.h) files for many of the common openPegasus classes (String, CIMParamValue, CIMObjectPath, etc.) are also heavily commented and contain example code.

FAQ 39 *Why do you place the brackets in your C++ programs in that strange way?*

Because placing brackets like this:

```
if (x == 6)
    {
    y = 9;
    z = 27;
    }
```

is both clearer for the reader and more logical than the repellent form

```
if (x == 6) {
    y = 9;
    z = 27;
}
```

A conditional statement in C++ can be followed by a single statement which, traditionally, is indented. If you want more than a single statement, then these can be made into a single group by the use of brackets. To be consistent, this group, including its brackets, should be indented.

Anyway, this is not a book about C++ style—just accept that the positions of the brackets in the examples are not accidental. The lack of religious debate on bracket positioning is one of the reasons why I like the Python language.

```
[Version("1.0.1"),
 Description(
        "ACNE_PBX is the logical representation "
        "of the whole model 34/AGB/76 PBX.") ]
class ACNE_PBX : CIM_ComputerSystem
     {
     // assume that there are no
     // properties other than those
     // inherited from CIM_ComputerSystem
     };

// ***********************************
// static instance of the ACNE_PBX
// created for test purposes
// ***********************************

instance of ACNE_PBX
     {
     CreationClassName = "ACNE_PBX";        // key
     Name = "XYZCompanyPBX1";               // key

     PrimaryOwnerName = "XYZCompany";
     PrimaryOwnerContact = "14 Cambridge St, Godmanchester";
     Roles = { "Primary PBX" };

     NameFormat = "Other";
     Dedicated = { 2 };                     // other
     OtherDedicatedDescriptions = { "PBX" };
     ResetCapability = 3;                   // may be reset
     };
```

Figure 12.3 *mof* **Code to Create Instance of ACNE_PBX**

```cpp
#include <Pegasus/Common/Config.h>
#include <Pegasus/Common/String.h>

#include "PbxTelephoneModule.h"

PEGASUS_NAMESPACE_BEGIN

extern "C" PEGASUS_EXPORT CIMProvider *
    PegasusCreateProvider(const String & name)
    {
    // if he's looking for one of our classes,
    // then return it. Otherwise return 0.

    if (String::equalNoCase(name,"PbxInstanceProvider"))
        return(new PbxInstanceProvider());
    if (String::equalNoCase(name, "PbxMethodProvider"))
        return(new PbxMethodProvider());

    return(0);
    }

PEGASUS_NAMESPACE_END
```

Figure 12.4   C++ Code to Create Providers: PbxMain.cpp

```
// *********************************************
// method   createInstance
// purpose  create a new PbxTelephoneModule
//          instance
// note     this operation is not allowed so
//          this method simply throws an
//          exception
// *********************************************

void PbxInstanceProvider::createInstance(
        const OperationContext & context,
        const CIMObjectPath & instanceReference,
        const CIMInstance & instanceObject,
        ObjectPathResponseHandler & handler)
    {
    // operator is not allowed to create
    // an instance: instances are created
    // implicitly by plugging the modules
    // into the shelf.

    throw CIMNotSupportedException(
            "Modules may not be created manually");

    return;
    }
```

Figure 12.5   C++ Code for CreateInstance()

```
// ********************************************
// method  enumerateInstanceNames
// purpose return names of all instances
// ********************************************

void PbxInstanceProvider::enumerateInstanceNames(
        const OperationContext & context,
        const CIMObjectPath & classReference,
        ObjectPathResponseHandler & handler)
    {
    // begin processing the request
    handler.processing();

    // pretend that we have scanned the hardware to
    // get the instances: actually just get them
    // from our table

    for (unsigned int i=0;i<numberInstances;i++)
        {
        // create an Object Path from the instances
        // that we have found from the hardware

        CIMObjectPath inst(instances[i]);

        // give the instance to the WBEM server

        handler.deliver(inst);
        }

    // complete processing the request
    handler.complete();
    return;
    }
```

Figure 12.6   C++ Code for EnumerateInstanceNames()

```
// ***********************************************
// method  buildInstance
// purpose build a dummy instance pretending
//          to read the real hardware
// ***********************************************

CIMInstance PbxInstanceProvider::buildInstance(
            String deviceId)
    {
    // create an instance of the correct class

    CIMInstance inst("ACNE_PBXTelephoneModule");

    // then add the properties

    inst.addProperty(CIMProperty("SystemCreationClassName",
                        String("ACNE_PBX")));
    inst.addProperty(CIMProperty("SystemName",
                        String("XYZCompanyPBX1")));
    inst.addProperty(CIMProperty("CreationClassName",
                        String("ACNE_PBXTelephoneModule")));
    inst.addProperty(CIMProperty("DeviceID",deviceId));
    inst.addProperty(CIMProperty("Protocol",1));
    inst.addProperty(CIMProperty("ModuleNumber",4));
    inst.addProperty(CIMProperty("Name",
                        String("Telephone Module")));
    return inst;
    }
```

Figure 12.7   Artificial C++ Code to Build a Dummy Instance

```cpp
// ****************************************************
// method    checkKeys
// purpose   check whether a requested instance
//           has a valid key
// output    throw exception if not valid, otherwise return
//           a string containing the deviceID
// ****************************************************

String PbxInstanceProvider::checkKeys(const CIMObjectPath & ref)
    {
    String  deviceId;
    CIMName keyName;
    int keyCount = 4;
    Array<CIMKeyBinding> keys = ref.getKeyBindings();

    if ((unsigned int)keys.size() != (unsigned int)keyCount)
        throw CIMInvalidParameterException("Must be 4 keys");

    for (unsigned int i = 0; i < keys.size(); i++)
        {
        keyName = keys[i].getName();
        if ((keyName.equal("DeviceID")) && (deviceId.size() == 0))
            {
            deviceId = keys[i].getValue();
            keyCount--;
            }
        else
            {
            if (keyName.equal("SystemCreationClassName") &&
                String::equal(keys[i].getValue(), "ACNE_PBX"))
                {
                keyCount--;
                }
            else
                .....etc for CreationClassName and SystemName.....
            }
        }
    if (keyCount != 0)
        throw CIMInvalidParameterException("Too few parameters");
    return deviceId;
    }
```

Figure 12.8   C++ Code for checkKeys()

```
// ***********************************************
// method getInstance
// purpose return a particular instance
// (if it exists)
// input
// ***********************************************

void PbxInstanceProvider::getInstance(
        const OperationContext & context,
        const CIMObjectPath & ref,
        const Boolean includeQualifiers,
        const Boolean includeClassOrigin,
        const CIMPropertyList & propertyList,
        InstanceResponseHandler & handler)
    {
    CIMInstance instance;
    String deviceId;

    // use the helper function to decide whether the
    // keys that we have been given are OK or not

    deviceId = checkKeys(ref);

    // now go away to get the instance requested (if it exists)
    // Of course, for this example it just returns an instance
    // read from a table but in the real world it would return
    // something genuine

    handler.processing();

    instance = buildInstance(deviceId);

    handler.deliver(instance);
    handler.complete();

    return;
    }
```

Figure 12.9   C++ Code for getInstance()

```
// *******************************************
// method  invokeMethod
// purpose look to see which method should
//         be invoked and invoke it
// *******************************************

void PbxMethodProvider::invokeMethod(
          const OperationContext & context,
          const CIMObjectPath & objectReference,
          const CIMName & methodName,
          const Array<CIMParamValue> & inParameters,
          MethodResultResponseHandler & handler)
   {
   // convert a fully qualified reference into a local reference
   // (class name and keys only).

   CIMObjectPath localReference = CIMObjectPath(
                           String(),
                           String(),
                           objectReference.getClassName(),
                           objectReference.getKeyBindings());

   handler.processing();

   if (objectReference.getClassName().equal(
                       "ACNE_PBXTelephoneModule"))
      {
      if (methodName.equal("setIndicator"))
         {
         Boolean retVal = setTheIndicator(localReference,
                              inParameters);
         handler.deliver(retVal);
         }
      if (methodName.equal("reset"))
         {
         Uint32 retVal = doReset(localReference);
         handler.deliver(retVal);
         }
      }

   handler.complete();
   }
```

Figure 12.10 C++ Code for InvokeMethod()

```
// **********************************************
// method   doReset
// purpose  pretend to do a RESET on the module
// input    instance Object Path
// output   0 if all OK, 2 otherwise
// **********************************************

Uint32 PbxMethodProvider::doReset(CIMObjectPath obj)
    {
    // OK, we'll pretend that we've done the reset

    return 0;
    }
```

Figure 12.11   C++ Code for doReset()

```
// *********************************************
// method   setTheIndicator
// purpose  pretend to set the indicator LED
// input    instance Object Path
//          input parameters
// output   previous value of the indicator
// *********************************************

Boolean PbxMethodProvider::setTheIndicator(CIMObjectPath obj,
        const Array<CIMParamValue> & inParameters)
    {
    CIMName keyName;
    String deviceId;

    // we must first retrieve the value of DeviceID
    Array<CIMKeyBinding> keys = obj.getKeyBindings();
    for (Uint32 i=0;i < keys.size();i++)
        {
        keyName = keys[i].getName();
        if (keyName.equal("DeviceID"))
            deviceId = keys[i].getValue();
        }
    if (deviceId.size() == 0)
        throw CIMInvalidParameterException("DeviceID missing");

    // and then get the parameter
    if (inParameters.size() != 1)
        throw CIMInvalidParameterException(
                    "setIndicator needs 1 param");
    CIMValue paramVal = inParameters[0].getValue();
    if (paramVal.getType() != CIMTYPE_BOOLEAN)
        throw CIMInvalidParameterException(
            "setIndicator param must be Boolean");
    Boolean inValue;
    paramVal.get(inValue);

    // send back the opposite value from the one input
    if (inValue == true)
        return false;
    else
        return true;
    }
```

Figure 12.12   C++ Code for setTheIndicator()

```
// ***********************************************
// create an instance of a PG_ProviderModule
// to group together a number of providers
// related to the PBX
// ***********************************************

instance of PG_ProviderModule
     {
     // description of the module
     Description = "PBXLogicalModule";

     // version of the module
     Version = "1.0.0";

     // version of the pegasus API definition
     InterfaceVersion = "2.1.0";

     // shared library name for the provider
     // module. e.g., Location = "x" means
     // that the shared library is libx.so
     Location = "PBXLogicalCode";

     // unique name for this module (not
     // a particular provider)
     Name = "PBXLogicalProviders";

     // name of the vendor producing this
     // module
     Vendor = "Chris and Ying Inc";

     // programming language: in this
     // case C++
     InterfaceType = "C++Default";
     };
```

Figure 12.13 *mof* **Code to Instantiate a PG_ProviderModule**

```
// **********************************************
// create an instance of a PG_Provider to
// define the instance and method providers
// within the providerModule
// **********************************************

instance of PG_Provider
    {
    // pointer back to the ProviderModule

    ProviderModuleName = "PBXLogicalProviders";

    // unique name for the instance provider
    // within the provider module

    Name = "PbxInstanceProvider";
    };

instance of PG_Provider
    {
    // pointer back to the ProviderModule

    ProviderModuleName = "PBXLogicalProviders";

    // unique name for the method provider
    // within the provider module

    Name = "PbxMethodProvider";
    };
```

Figure 12.14 *mof* Code to Instantiate a PG_Provider

```
// ************************************************
// create an instance of PG_ProviderCapabilities
// to define precisely what our instance
// provider can do
// ************************************************

instance of PG_ProviderCapabilities
    {
    // pointer back to the ProviderModule
    ProviderModuleName = "PBXLogicalProviders";

    // pointer back to the ProviderName
    ProviderName = "PbxInstanceProvider";

    // which mof class does this provider support?
    ClassName = "ACNE_PBXTelephoneModule";

    // which functions does the provider provide?
    // (i.e. is it an instance provider or
    // method provider, etc?). Options are:
    // Instance (2)   Association (3)
    // Indication (4) Method (5)
    ProviderType = { 2 };

    // the namespaces within which this provider
    // functions
    NameSpaces = { "root/acnePbx" };

    // a name for this set of capabilities which is
    // unique within provider
    CapabilityId = "1";
    };
```

Figure 12.15   *mof* **Code to Instantiate PG_ProviderCapabilities for Instance Provider**

```
// **********************************************
// create an instance of PG_ProviderCapabilities
// to define precisely what our method
// provider can do
// **********************************************

instance of PG_ProviderCapabilities
    {
    // pointer back to the ProviderModule
    ProviderModuleName = "PBXLogicalProviders";

    // pointer back to the ProviderName
    ProviderName = "PbxMethodProvider";

    // which mof class does this provider support?
    ClassName = "ACNE_PBXTelephoneModule";

    // functions as in previous example
    ProviderType = { 5 };

    // the namespaces within which this provider
    // functions
    NameSpaces = { "root/acnePbx" };

    // a name for this set of capabilities which is
    // unique within provider
    CapabilityId = "2";
    };
```

Figure 12.16 *mof* **Code to Instantiate PG_ProviderCapabilities for Method Provider**

```
<?xml version="1.0" encoding="utf-8"?>
<CIM CIMVERSION="2.0" DTDVERSION="2.0">
  <MESSAGE ID="53044" PROTOCOLVERSION="1.0">
    <SIMPLEREQ>
      <IMETHODCALL NAME="CreateInstance">
        <LOCALNAMESPACEPATH>
          <NAMESPACE NAME="root"/>
          <NAMESPACE NAME="acnePbx"/>
        </LOCALNAMESPACEPATH>
        <IPARAMVALUE NAME="NewInstance">
          <INSTANCE CLASSNAME="ACNE_PBXTelephoneModule">
           <PROPERTY NAME="Protocol" TYPE="uint32">
            <VALUE>0</VALUE>
           </PROPERTY>
           <PROPERTY NAME="ModuleNumber" TYPE="uint16">
            <VALUE>4</VALUE>
           </PROPERTY>
           <PROPERTY NAME="SystemCreationClassName"
                                          TYPE="string">
            <VALUE>ACNE_PBX</VALUE>
           </PROPERTY>
           <PROPERTY NAME="SystemName" TYPE="string">
            <VALUE>XYZCompanyPBX1</VALUE>
           </PROPERTY>
           <PROPERTY NAME="CreationClassName" TYPE="string">
            <VALUE>ACNE_PBXTelephoneModule</VALUE>
           </PROPERTY>
           <PROPERTY NAME="DeviceID" TYPE="string">
            <VALUE>123</VALUE>
           </PROPERTY>
          </INSTANCE>
        </IPARAMVALUE>
      </IMETHODCALL>
    </SIMPLEREQ>
  </MESSAGE>
</CIM>
```

Figure 12.17 XML to Create an ACNE_PBXTelephoneModule

Chapter 13

Writing Clients and Listeners

What Clients Are Not

Before talking about clients and their architectures and implementations, it is important to be clear that WBEM clients are *not* user interfaces. As we have seen before, the chain of components is as follows:

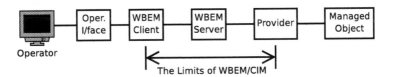

The WBEM client represents the operator in negotiations with the WBEM server; it converts the operator's requests into CIM-XML from whatever language is natural for the operator and it translates the WBEM server's response back. It is not responsible for producing drop-down menus on an operator's workstation, rendering HTML onto a screen, or parsing the text of a command line interface; that is the function of the operator interface software. Of course, the operator interface and WBEM client may be (and often are) implemented in the same piece of code.

We can therefore expect a WBEM client to be a sort of Janus with a relatively simple and stable interface towards the WBEM server and a much more complex and variable interface towards the operator. It is therefore impossible to discuss the WBEM client in isolation—most of its functionality deals with operator interfaces rather than WBEM. I will therefore spend some time in this chapter looking at possible WBEM client implementations for different types of operator interface.

One important architectural characteristic to remember when designing a WBEM client (or, indeed a provider) is that the client should not know about the providers with which it indirectly interacts—the WBEM server acts as a broker between the two and no knowledge of a particular provider structure should ever be built into a WBEM client.

Semantic Knowledge

Apart from the obvious matter of translation between interfaces, the major problem facing the WBEM client or Operator Interface, or both, is obtaining the semantic knowledge needed to ask the right questions of the WBEM server. The WBEM client knows, for example, how to parcel up a `getInstance()` request in XML and pass this to the WBEM server; that is easy. However, something needs to have the intelligence to change an operator's request for the number of users currently logged on into a call to `getInstance()` on the CIM_OperatingSystem class and the extraction of the `NumberOfUsers` property. This is a particularly simple example because only one property is involved. An operator request could reasonably involve accessing instances of several classes, traversing an association or two and then combining the values of several properties.

One major advantage promised by a unifying technology such as WBEM/CIM is that, if you buy and manage equipment from companies A and B, both of whom use the standard CIM models on their equipment, then the issue of semantic knowledge only needs to be addressed once—solve it for company A's equipment and you have solved it for company B's.

There are really only three places where this semantic knowledge (sometimes known as "business logic") can reside: in the head of the operator, in the operator interface or in the WBEM client.

I will deal with the first of these here, because it is simple, and defer consideration of the other two until later in this chapter. If the semantic knowledge can be assumed to be in the head of the operator, then a completely generic WBEM client can be built. The operator instructs the WBEM client to invoke the function `getInstance()` and retrieve the `NumberOfUsers` property for a particular namespace and the WBEM client does exactly that. There are several such generic clients available, mainly for debugging—they allow you to debug your model and providers in a controlled manner. Some of these generic clients are described on page 273, and as they are not really applicable for customer deployment, I will not consider them further.

Server-Side Client Implementation

The implementation of the WBEM server side of a particular WBEM client is conceptually very simple and the available WBEM server implementations provide code libraries to make the coding very straightforward. These libraries typically provide function calls to connect to and disconnect from the WBEM server and calls for each of the intrinsic functions. Internally, the code generates the necessary CIM-XML, sends it to the WBEM server, receives back the result, de-XMLs it and passes it back to the invoking program. This is all very simple.

Your code does, however, need some understanding of what is happening under the covers because this can seriously affect performance of the system. Because of the single exchange/response nature of HTTP, the basic operation of the WBEM client code when invoked can be summarised as follows:

1. Discover the whereabouts of the appropriate WBEM server. This is described in more detail below.
2. Connect to the WBEM server, carrying out whatever authentication is required. This is described in more detail on page 142 and in Chapter 7.
3. Issue a request and wait for the reply. This is described in more detail in Chapter 7.
4. Process the response by passing it to the operator or higher-level management system.
5. Repeat from step 2.

Because of the single request/response nature of HTTP, it is necessary to repeat the authentication step for each request.

This sequence can be varied slightly by using the HTTPS protocol where authentication can be carried out once and multiple commands be issued.

Discovery

This section deals with four problems facing a WBEM client:

1. Where is the WBEM server? There may be several WBEM servers in a network, each handling different namespaces. How does a WBEM client find the one with which it should communicate?

2. What can each WBEM server do? Not all WBEM servers support all CIM operations—perhaps a particular server cannot handle queries. Having found the WBEM server, how can the WBEM client determine its capabilities?

3. Having found the WBEM server and its capabilities, what namespaces does it support?

4. How many tasks has the WBEM server handled and how much information has it received and generated?

These questions are answered in the subsections below.

Finding the WBEM Server

At the time of writing, the task of finding a particular WBEM server (i.e., a WBEM server supporting a particular namespace) is being addressed by the DMTF's InterOp Working Group. It appears that the recommendation, when it emerges, will use the IETF's Service Location Protocol (SLP) and several WBEM server implementations, including openPegasus, have produced code based on that assumption.

SLP is described in RFC2608 and is effectively a protocol whereby so-called User Agents, acting on behalf of clients, can search for a particular service. This can be done in a small network by multicasting or broadcasting requests and in a larger network by making use of a Directory Agent. If you are familiar with CORBA, it might be useful to think of the Directory Agent as a Trader Service, receiving and storing registrations by servers and responding to requests from clients seeking a particular service.

WBEM servers register themselves with a Directory Agent or, in smaller networks, respond to multicast requests from clients.

Finding the Capabilities of the WBEM Server

Anticipating the problem of finding what operations a particular WBEM server supports, the DMTF has defined several classes which all WBEM servers must implement. These classes, which qualify for Olympic honours in any name length competition, are described in detail in the DMTF's "CIM InterOp Model" white paper (DSP0153) and are illustrated in Figure 13.1.

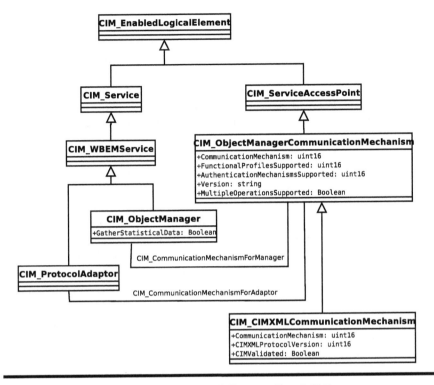

Figure 13.1   WBEM Server Capabilities

These classes inherit from the CIM_Service and CIM_ServiceAccess-Point classes described on page 92. If you are hazy on these two important classes, then I strongly recommend that you reread that section.

As Figure 13.1 illustrates, a CIM_ObjectManager *is a* CIM_WBEM-Service which *is a* CIM_Service.   Most of the useful information about accessing the WBEM server is held in an instance of the CIM_ObjectManagerCommunicationMechanism class which can be reached through an association on the CIM_ObjectManager instance. CIM_ObjectManager itself *does* hold one piece of information: a flag to specify whether statistical information about the WBEM server itself is being collected. If this flag is *false*, then information is not being collected; even if instances of the CIM_CIMOMStatisticalData class are present, no statistics are being gathered.

The instances of the CIM_ObjectManagerCommunicationMech-anism class are, however, the main source of information. They contain information about how the WBEM server may be accessed:

CommunicationMechanism. This is an enumeration describing an encoding for CIM messages and a protocol which may be used to access the WBEM server. Currently only CIM-XML and CIM-SOAP are defined by the DMTF, but you may define your own protocol and specify your support for it here.

If your WBEM server supports multiple protocols, then one instance of CIM_ObjectManagerCommunicationMechanism will have been created for each protocol.

FunctionalProfilesSupported. For the purposes of determining what a particular WBEM server can do, the intrinsic methods are grouped into so-called functional groups. These groups are interrelated in the way illustrated in Figure 13.2. Functional-ProfilesSupported is an array of enumerated values which specifies which functional groups the WBEM server supports. Of course, all WBEM servers must support Basic Read operations, otherwise the WBEM client could not access the instances of this class to find out that the CIM server cannot do Basic Read operations!

The basic classifications given in Figure 13.2 can be broken down further by using free-form strings to specify, for example, precisely *which* query languages are supported.

You need to be careful when interpreting the Functional-ProfilesSupported enumeration. If a particular WBEM server indicates that it does not support a particular functional group, then you can be sure that it does not support any of the operations in that group. If, on the other hand, a WBEM server indicates that it *does* support a particular group then this only indicates that zero (*sic!*) or more of the operations in that group are supported—i.e., it tells you nothing. If you try to invoke a function which is not supported then the WBEM server will return the CIM_ERR_NOT_SUPPORTED exception.

MultipleOperationsSupported. Some WBEM servers can only handle a single request in a message, others can handle several—this Boolean property specifies this. As we will see later, the WBEM server keeps statistics about the number of times each intrinsic method has been called. If multiple operations are included in a single call, then these counters are not kept for individual operations—a single "multiple operation" counter being incremented instead.

AuthenticationMechanismsSupported. This defines the types of authentication in which the WBEM server is willing to participate. See page 142 for more details.

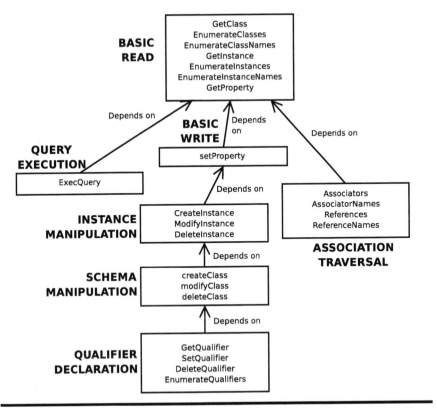

Figure 13.2 Functional Groupings of Intrinsic Methods

Version. The version number of the protocol in the form M.N where M is the major version number and N the minor one.

Finding the Supported Namespaces

The InterOp common model from the DMTF also covers the question of determining the namespaces supported by a WBEM server. The CIM_ObjectManager class described above has an additional association, CIM_NamespaceInManager, which associates it with a CIM_Namespace class.

Namespaces can therefore be created within a WBEM server by creating an instance of the CIM_Namespace class and associating it with the CIM_ObjectManager. Namespaces can be found by following the CIM_NamespaceInManager associations from the CIM_ObjectManager instance.

Getting Statistics from the WBEM Server

If the CIMOM is collecting statistical data (see the GatherStatisti-calData property of CIM_ObjectManager), then one instance of the CIM_CIMOMStatisticalData class is created for each type of intrinsic method call handled by the WBEM server.

For example, a particular instance can be read by a WBEM client to determine how many calls have been made to `getInstance()`, how much time was spent in total in the WBEM server during these calls, how much time was spent in providers during these calls and the total size of the responses sent back as a result of these calls.

As you can imagine collecting these data can consume a lot of processor time and their collection should be turned on circumspectly.

Operator-Side Client Implementation

Strictly speaking, this interface has nothing to do with WBEM/CIM and is therefore not appropriate for this book. In the same way that you can get a true musician out of bed in the morning by playing almost a whole scale on a piano (he or she has to get out to complete it; otherwise it is unbearable), so I feel that to have covered almost the whole chain illustrated on page 247, and not having reached the operator would be cruel to my readers.

I said earlier that the semantic knowledge (business knowledge) can reside in the mind of the operator, in the operator interface code or in the WBEM client. I dealt briefly with the first of these, resulting in a generic WBEM client and will consider the other two here.

Some operator interfaces are complex, particularly those which display graphical user interfaces and, to be tractable, they need to be at least partially table driven; otherwise they will collapse under the weight of too many *ad hoc* rules. Other operator interfaces may be very simple. A command-line interface (CLI), for example, may contain code to check the syntax of what the operator types but is generally unconcerned with the semantics of what is typed.

Some of the possible operator interface/WBEM client combinations are shown in Figure 13.3 with arrows to show where semantic intelligence could reside. Putting this intelligence other than in the interface itself is sometimes called creating a "thin client," and not surprisingly, an interface with this intelligence is called a "thick client" (this is the usage of "thick" to mean wide rather than stupid). Note that "client" in this context does not mean a WBEM client—indeed, if the client is thin, then the WBEM client is likely to be thick and *vice versa*.

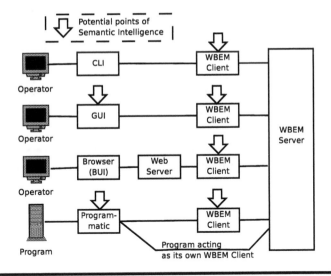

Figure 13.3   WBEM Clients

A browser, by its nature, is a thin client because it only has the ability to take HTML and render it onto a screen.

Of the operator interface/WBEM client permutations illustrated in Figure 13.3, the one associated with the programmatic interface (i.e., a higher-level management system) is likely to be the simplest, since the higher-level system is likely to have the semantic intelligence available to it. In some cases, it is likely that the higher-level management system will be its own WBEM client (as illustrated in Figure 13.3).

Browser user interfaces (BUIs) are becoming increasingly popular, particularly among small devices, such as IP routers intended for the home. To use a WBEM client with such an interface, the technique is to install a Web server (e.g., Apache) and code the WBEM client as a CGI program. If you are unfamiliar with CGI programs then see its entry in Appendix G, the Glossary.

Providing the semantic intelligence is difficult if you are trying to produce a purely generic WBEM client which will be able to manage my toaster today and an IP router tomorrow. Some help is forthcoming from the choice of WBEM/CIM because intrinsic commands exist, not only to access instances from the WBEM server, but also to access classes and the structure of the data. This still does not give quite enough information to allow the WBEM client to be completely generic; although it can at least determine that a property `NumberOfUsers` exists, it still cannot automatically determine that it must translate an operator request for the current number of users to

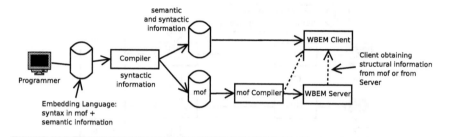

Figure 13.4   Embedding *mof* into a larger language

the call `getInstance()` on the CIM_OperatingSystem class and the extraction of the `NumberOfUsers` property.

It would, however, be possible to define a new qualifier, say `ACNE_COMMAND`, and to add this to the `NumberOfUsers` property:

```
[ACNE_COMMAND("Show Number Users"),
 Description (
    "Number of user sessions for which the OperatingSystem "
    "is currently storing state information."),
    Gauge,
    MappingStrings {"MIF.DMTF|Host System|001.4",
        "MIB.IETF|HOST-RESOURCES-MIB.hrSystemNumUsers"} ]
    uint32 NumberOfUsers;
```

This would allow the WBEM client to match the operator command "Show Number Users" with this property as it scanned the schema.

On the positive side, this type of technique would allow a completely generic client to be built and, for a small system, it might (just) be practical. Realistically, it would very quickly become unwieldy, particularly if extended to include commands which needed to access combinations of properties and classes.

This technique can, however, be extended either by embedding a semantic language into *mof* specifications by using specially formatted comments which are ignored by the WBEM server but which could be interpreted by the WBEM client, or by embedding the *mof* into a wider language and extracting it separately as shown in Figure 13.4.

Before you go to the extreme of defining new languages and writing new compilers, you should ask yourself how generic your WBEM client and operator interface need to be. If your aim is to produce a particular management interface to a particular device (rather than a generic management interface to any device), then it might be acceptable to hard code the more stable parts of the models into the code.

The CIM_OperatingSystem class has existed in the System common model for many releases and is likely to remain there for many releases to come. Would it be possible to code into your WBEM client program the knowledge that, when the operator asks for the number of users, then it should check that an instance of CIM_OperatingSystem exists, obtain it and extract the **NumberOfUsers** property? In many instances this hard coding is acceptable, particularly if it is done in such a way that it can be easily updated.

This technique is less risky than it might at first appear. The CIM_NetworkPort class, for example, has certain properties and associations. If your company has introduced a new type of interface port, its class is likely to have been created as a child of CIM_Network-Port and all of the functions that you have already functioning with other children of CIM_NetworkPort are likely to work for your new port type. To some extent, the core and common models put model designers into a straitjacket and this makes the external management much easier.

Frequently Asked Questions

FAQ 40 *What types of operator interface (Command-Line Interface, graphical interface, etc.) does CIM support?*

None. CIM does not concern itself with the details of how the information is displayed to the operator. The role of CIM is to organise the information relating to a device or service in a coherent and structured manner and allow the WBEM client, working on behalf of the operator, access to it. This is completely independent of the way in which the retrieved information is displayed on the operator's screen. In some cases, for example when a program is configuring multiple devices, there may be neither operator nor screen.

FAQ 41 *How can a client determine precisely what commands a WBEM server supports?*

By reading the instances of the CIM_ObjectManager, CIM_ObjectManagerCommunicationMechanism, and CIM_CIMXMLCommunicationMechanism classes as described above. However, if a WBEM server claims to be able to support a particular functional group then each should be tested to see whether a CIM_ERR_NOT_SUPPORTED exception is returned.

Chapter 14

Transition to WBEM/CIM

If your company manufactures a range of products with a well-proven, proprietary management interface, provided by software which has reached its steady-state maintenance mode, and if there is no discernible customer pressure to change, then this chapter is not for you.

If you work for a new company without shipped product, and have decided that your first released product will be based on WBEM/CIM, then this chapter is not for you either.

Instead, I am writing here for engineers and programmers in the tens of thousands of companies with shipped product, the management of which is based on proprietary protocols or SNMP or a mixture of both. This installed base cannot be discarded overnight but there is mounting customer pressure for service-level management and alignment of management interfaces with your competitor's product. And, of course, your competitor has used WBEM/CIM. There may also be internal pressure to reduce the maintenance cost of the existing management system.

There are many ways to design a system and I shall do no more in this chapter than outline a few. A tool that will appear more than once is the use of the `MappingStrings` qualifier; I describe this important qualifier in appendix E.

The next few sections describe practical architectures, but there is something of an underlying, theoretical structure beneath them. This is covered in part in Chapter 6 and Chapter 7 of the CIM specification (DSP0004) and I touch on it briefly on page 265. If you are familiar with this theory, then you will realise that I am applying the so-called "domain" interoperability technique.

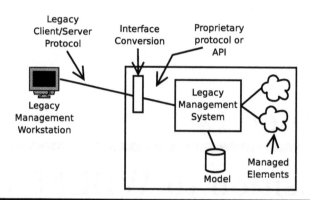

Figure 14.1   A Legacy Management System

Some Upgrade Architectures

I assume that Figure 14.1 encompasses the main points of your legacy system: some form of non-WBEM protocol is used between management workstations and the device being managed. A protocol conversion is then carried out (which may be as simple as removing transport protocol headers) before the messages are passed to the on-device management software. This software has some model of the device (possibly implicit in a long C++ **case** statement, perhaps an SNMP MIB) which it uses to respond to the operator's command. Figure 14.1 could, of course, with a little relabelling, be a picture of WBEM/CIM or, indeed, almost any other management system.

The question then is, "What do you want to do with this system?" I consider here the following scenarios. Although these cannot cover all possibilities, I have chosen them so that their components are widely applicable.

> **Scenario 1.** The on-device code is ancient, poorly documented and the person who wrote it has left the company. It cannot be touched without a major investment in reverse engineering, rework and retest. Customers are demanding service-level management and the on-device code makes it difficult to manage the new generation of cards which your hardware development team is currently designing. I am sure that this scenario is unfamiliar to you—assume that I have included it for other readers.

> **Scenario 2.** The off-device management system is SNMP-aware and is being used to capture and display SNMP traps (i.e., alarms). Configuration management is currently done through a command-line interface that may be enhanced but not re-

placed because it would break many tools which have been developed over the years for automated management. While retaining these two interfaces, there is also industry pressure to provide an additional, WBEM-based interface. The intention is that the legacy system will be capped and, over time, evolved to WBEM/CIM.

Scenario 3. The code which actually interfaces to the hardware of the device is supplied by the hardware manufacturer and source code is not made available. Although it is intended to move to a full WBEM/CIM-based management system, alarms from these devices will still be raised as SNMP traps.

Scenario 1: Untouchable Existing System

This situation appears to be very common, certainly in the telecommunications industry. My approach here is to encapsulate and freeze the existing system and gradually evolve it, subsystem by subsystem, to the new world, starting with the more complex subsystems.

I have illustrated the first step in this in Figure 14.2: a WBEM server has been introduced on the device and a CIM model founded on the DMTF's core and common models has been designed, covering not only existing management functions but also the new service management and any other device management which the legacy system doesn't cover. In effect, we are going to use the legacy management system as a large and opaque CIM provider for handling the low-level device management, with a small protocol conversion provider put in front of it to support the legacy protocol or API.

Unless the existing on-device management system is completely *ad hoc*, it is likely to have some mechanism for addressing components uniquely (the equivalent of CIM's object path or SNMP's OID). If this mechanism is present, then we can make use of the mapping string concept to simplify the interface between the new provider and the old management system. Note that, although the normal use of mapping strings is to map CIM properties to international standards, this is not necessary—you can use them to map CIM properties to your own object identifiers.

An operator command then arrives at the WBEM server, couched purely in terms of the CIM model, providing the management compatibility with your competitors that your customers are demanding. For elements supported by the legacy system, the WBEM server passes the command to the appropriate provider which accesses the mapping strings. If the original interface was symmetric and clean, the presence of the mapping strings allows the provider to be coded in a reasonably

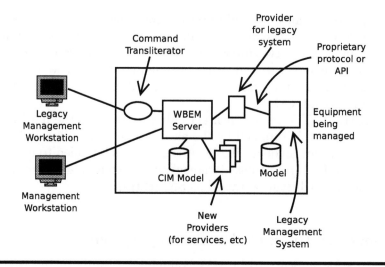

Figure 14.2 Encapsulation of a Legacy System

generic manner. If the original interface was not clean, then even with the mapping strings, you may have to write some nasty code. In either case, the provider can translate the command into a form suitable for the legacy system.

This encapsulation means that the legacy system can initially remain unmodified; it need have no knowledge of the presence of the WBEM server or the service-level commands. Similarly, the WBEM server has no special knowledge either—it simply sees a provider with a standard interface. This technique has drawbacks, primarily the amount of memory required to hold both management systems and the degradation of performance caused by low-level commands having to pass through two management systems.

There is also a psychological disadvantage: once the system is working in this manner there is less incentive to complete the migration and thereby simplify the software structure. Even if this type of migration project is started with the best intentions of returning to complete the transition, once the system is working it is often difficult to justify spending R&D money on software development that does not deliver customer-visible features.

Once the encapsulation is in place, migration takes place in several steps:

■ New features added to the device are managed only by the WBEM server, the CIM model being enhanced and the providers

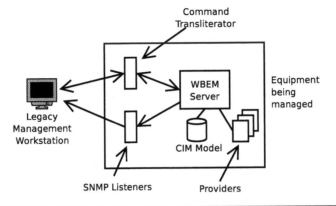

Figure 14.3   Supporting a Legacy Interface

being written as required. This caps the growth of the legacy system.

■ Any service-level features currently supported by the legacy system are remodelled in CIM and providers are written, removing this level of abstraction from the legacy system. I have found that this often comes as a relief to the maintainers of the legacy system, these higher-level features typically having been added in a somewhat *ad hoc* manner to a perfectly good device management system. Because of the impedance mismatch between the two parts, bugs in the legacy system tend to occur either in the service-level code or in the interface between service and device management. Removing the service-level management allows the legacy system to revert to its previous, more stable state.

■ If it felt useful, device management features of the legacy system are moved to the WBEM server in an orderly fashion, possibly as the hardware of a board is replaced or a software feature rewritten, necessitating new management software.

Scenario 2: Preserving Operator Interfaces

This problem is to some extent the reverse of scenario 1: you need to preserve the external interface but are allowed to change the on-device code. In some ways this is a harder problem, particularly if the schema-modifying commands supported by WBEM/CIM are new to the system and need to be supported. There are, for example, no equivalents in SNMP to `CreateClass()`, `DeleteClass()`, etc.

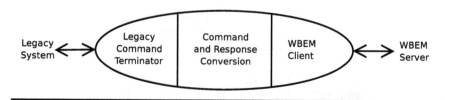

Figure 14.4   Structure of the Command Transliterator

My recommended approach is illustrated in Figure 14.3; the legacy operator interface is treated no differently than any other operator interface: a command transliterator is required with knowledge of the legacy interface and of the conversion between the operator interface and the WBEM/CIM world. This command transliterator acts as a WBEM client to the WBEM server and as a server to the legacy management workstation application. It can be considered to have three parts; see Figure 14.4. The legacy command terminator checks the syntax of the incoming command, responds to local commands such as requests for help, and formats responses back to the legacy system. For many popular interfaces (e.g., Cisco IOS interface) there are freely available software modules for this function. The other side of Figure 14.4 is the WBEM client. Again, this software is freely available from any WBEM server source. The work involved in creating the transliterator is therefore the conversion of the commands and responses between the legacy and WBEM worlds.

This conversion requires syntactical knowledge of the CIM interface and semantic knowledge of the structure of the information in the CIM model. In this scenario we are helped in obtaining the semantic knowledge, because some of the SNMP objects are already modelled in the DMTF core and common models and have mapping strings already in place. The problem, however, is that mapping strings are not keys. Given a particular SNMP object reference, there is no efficient way in which the corresponding CIM class can be retrieved from the WBEM server. There are two obvious solutions to this problem:

■ Allow the conversion routine to walk the entire CIM model (i.e., enquire about every class) when the program is first started. This may be time consuming but is only needed once if it uses the information to build an internal table, indexed by an SNMP object identifier.

■ Provide the conversion routine, again at system initialisation, with a copy of the schema information—possibly by writing an additional back-end to the *mof* compiler.

Both of these techniques will, of course, only work if the schemata are stable between software releases—any changes made to the structure of the schemata themselves by operators (through the use of `addClass()`, etc.) will not be reflected in the copy held by the conversion routine. In most transition environments this is not a problem since the legacy workstation software typically cannot support operator commands for schema changes.

Another decision which you will need to make if you adopt this approach is where to run the command transliterator. It might be more appropriate to combine it with the workstation software than to place it on the managed device.

Indications arising on the device can be routed to a listener which can convert them to SNMP traps.

Scenario 3: Preserving On-Device Drivers

This is probably the simplest of the three scenarios: a provider can hide the actual configuration interface for the hardware devices from the WBEM server.

The traps emanating from the hardware can be caught by a indication provider that creates an instance of a class derived from CIM_SNMPTrapIndication—see Figure 8.3 on page 155.

Once this has passed through the WBEM server, it can be directed to listeners specified in instances of the CIM_SNMPTrapTarget class—see Figure 6.6 on page 104.

This mapping of SNMP trap fields to CIM properties was foreseen by the DMTF and is defined cleanly in the various classes.

Some Theoretical Background

Sections 6 and 7 of the CIM Specification (DSP0004) discuss various ways to map between modelling paradigms. In particular they describe the technique, recast and domain approaches. I have illustrated these in Figure 14.5 which I have modified slightly from the paper "Semantic Management: Advantages of Using an Ontology-Based Management Information Meta-Model" by de Vergara, Villagrá, and Berrocal.

In Figure 14.5, "source domain" might be your legacy management system which you wish to map into CIM, the "destination domain." This would reflect scenario 1. Alternatively, CIM could be the "source domain" and your legacy system the "destination domain"—as in scenario 2.

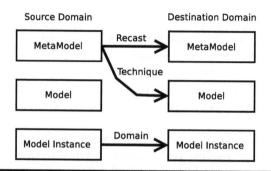

Figure 14.5   Different Mappings

I mentioned at the beginning of Chapter 6 that, in addition to the core and common models, the DMTF has also produced a meta-model. This model describes the modelling terms themselves—the concept of a class, the fact that classes can have properties and methods, etc. If you find the concept of a meta-model a little abstract, then you should be aware that the Object Management Group (OMG) defines *four* layers of models in its meta-object facility (another *mof*!) specifications: the information (instances, in our terms), the model which describes how instances may be created, the meta-model which describes how models may be created, and the meta-meta-model which describes how meta-models may be created.

I assume that, whatever legacy management system you are using, it will have some form of implicit or explicit meta-model. In many cases of proprietary management systems, this model was never made explicit but it exists in the minds of the designers. Given this, the three basic techniques are as follows:

1. **Recast mapping:** If it is feasible, this is the cleanest way of providing a mapping. The manual work involved is to match meta-constructs in each modelling language and write a compiler to convert one language to the other. For example, it might be possible to say that *attributes* in your legacy system correspond exactly with *properties* in CIM, that *components* in the legacy system correspond to *classes* in CIM, etc. If such a mapping can be defined, then conversion of one model into the other can be carried out in an automated manner. This technique has been used successfully to inter-translate SNMP MIBs, CORBA IDLs, and TMN MIBs (see "Inter-Domain Management: Specification and Interaction Translation," published by the Open Group as document C802).

2. **Technique mapping:** As Figure 14.5 shows, this is a cross-layer approach. The idea is to take concepts in the legacy meta-model (e.g., an attribute) and map these not to the destination meta-model but to the destination model (e.g., classes in CIM). This idea was used many years ago in the DMTF's standard DSP0002 published in 1997 to define a mapping between SNMP and the desktop management interface (DMI). Although the document is now only of archeological interest, the technique it used is still relevant.

   Sun Microsystems' `mib2mof` utility uses this technique to translate Solaris SNMP MIBs into *mof* files.

3. **Domain mapping:** This is the most labour-intensive technique and it commonly results in an incomplete mapping. It can, however, be used in circumstances where recast is not applicable. Domain mapping is a content-to-content mapping: mapping, for example, the *LocalDateTime* property of the CIM_OperatingSystem class to the *HOST-RESOURCES-MIB.hrSystemDate* field of the IETF's SNMP MIB. The equivalence is indicated by the MappingStrings qualifier.

   Because there is no attempt to map the meta-model, semantics are lost in the mapping: the field may be equivalent in the syntactical sense but is it actually used in the same way in both the source and the destination models?

Generally, with the possible exception of those systems where a recast mapping is possible, semantic information is lost in any translation: between SNMP and CIM, for example, hierarchical information must change its meaning from *has a* to *is a*. This means that some process with semantic intelligence must be present, possibly a provider or client interface, to make the necessary semantic transformations.

There is still much research being carried out on the automated, lossless conversion between management models. The state of the practical art today is that semantic information will almost certainly be lost in anything other than transformations between systems with isomorphic meta-models.

Chapter 15

Implementations and Tools

Figure 4.4 on page 39 illustrates the components of a WBEM server:

- A CIMOM, together with its peripheral interfaces, capable of accepting requests from clients, accessing the repository and providers and composing responses.
- A repository—effectively a database—to hold the structural information about the model.
- A compiler capable of checking the syntax, compiling models written in *mof* and acting as a client to load the model into the CIMOM.

We would expect any implementation to provide these basic elements but the following would be also useful:

- Test providers and clients to act as exemplars for engineers writing new ones and also to act as regression test tools.
- Generic graphical, browser, or command-line clients to allow the repository to be browsed and providers to be invoked in a controlled manner.
- Generic test clients which can be used for the regression testing of the WBEM server and providers.
- Providers for common components: Linux or Windows operating systems, etc.
- *mof* editors and UML drawing tools.
- Code generators to produce skeleton C, C++, or Java code automatically from the *mof* definitions.

Documentation and support are also, of course, essential.

The next few sections give a brief summary of a few of the WBEM servers, WBEM toolsets, and generic clients that are available on the market. There are dozens of WBEM server implementations and my list is certainly not complete—it reflects the bias of my experience.

There is an open source movement called the "WBEMsource Initiative" which is an umbrella organisation founded to coördinate open source WBEM projects to ensure interoperability and portability between them. Several of the open source WBEM servers, including open-Pegasus, OpenWBEM, and WBEM Services, are part of this organisation. For more details, see `http://www.wbemsource.org/`.

WBEM Servers

openPegasus

openPegasus is an open source, C++-based WBEM server created by individuals and programmers committed by a number of companies (including IBM, HP, and EMC) under the umbrella of the Open Group (see `http://www.opengroup.org`). The code is released under an MIT-style Source Licence which I include in Appendix H and is available from the openPegasus Web site, `http://www.openPegasus.org`, in source code form as a "tar-ball" snapshot or as an extraction from CVS and in prebuilt form for a number of platforms. Because this is the WBEM server which I use for the example in Chapter 12, I have given full installation instructions in Appendix F.

Given its joint development by a number of companies, the platforms on which it runs are naturally wide and include:

- UNIX (AIX, HPUX, Solaris, Tru-UNIX)
- Linux
- OS 400
- Z/OS
- OpenVMS (alpha and IA64)
- Microsoft Windows (NT, 2000, 9x)

For many embedded applications, one possible criticism of the open-Pegasus CIMOM is its size—it runs primarily on large servers where memory and disk footprint are not significant. There has been discussion within the openPegasus community for some time about producing a "Pegasus Lite" with a much smaller memory footprint but, at the time of writing, nothing is apparently happening in this area.

Several large companies are using the openPegasus code in their products which argues for its stability, level of support and readiness for deployment.

OpenWBEM

OpenWBEM is another C++-based, open source WBEM server, created initially by Caldera and now maintained by Vintela Inc (`http://www.vintela.com`). The code is available from the openWBEM Web site, `http://www.openwbem.org`.

In many ways the architecture of OpenWBEM is much cleaner than that of openPegasus and providers are easier to write. OpenWBEM, however, has fewer companies driving its development—releases and innovations are slower and fewer than openPegasus. Because the DMTF standards are evolving, the slow innovation rate of OpenWBEM could also mean that it is diverging from the standards.

WBEM Services

This is an open source, Java-based WBEM server and is probably the most complete of all of the open source projects. Certainly the documentation is many times better than that of openPegasus and Open-WBEM. The project is supported strongly by Sun Microsystems and details can be found at `http://wbemservices.sourceforge.net/`. The code is released under Sun's "Sun Industry Standards Source License" (SISSL).

SNIA Open Source CIMOM

The Storage Network Industry Association's (SNIA's) CIMOM (actually a full WBEM server—the developers are currently seeking a new name for the product) is written in Java and, like openPegasus, is accessible through the Open Group Web site (`http://www.opengroup.org/snia-cimom/`).

This WBEM server is very complete and forms the basis of SNIA's Storage Management Initiative Specification (SMI-S) (previously known as Bluefin) which was introduced in mid-2002 to encourage the industrywide adoption of an open interface for the management of storage networks. SMI-S is the basis of regular interoperability tests, known as "plug fests," organised by SNIA at which dozens of vendors of storage equipment demonstrate the inter-operability of their management systems.

The code is currently released under yet another open source licence: the Storage Networking Industry Association Public Licence, version 1.1. There is intent within the development community to migrate to the same licence as that used by openPegasus.

J WBEM

WBEM Solutions Inc sells a Java-based WBEM server known as J WBEM. This is a full implementation of the DMTF standards and WBEM Solutions engineers are active in the DMTF standards development. A C-based WBEM server, known as C WBEM, was released in the fourth quarter of 2003, in particular to address the market for embedded servers where memory footprint is particularly important. For more details, see `http://www.wbemsolutions.com/index.html`.

WMI

Although Microsoft ships a WBEM server, known as WMI, with copies of its Windows operating systems, the client/server interface is based on Microsoft's own COM/DCOM technology and Microsoft does not make the source code of the WBEM server publicly available.

The primary purpose of this WBEM server is to simplify the management of the Microsoft Windows operating system, but it can, of course, be used to manage other parts of a system running on Windows.

Solaris WBEM

The source code for this WBEM server has been released by Sun Microsystems as part of the WBEM Services initiative described earlier—the code is available at `http://wbemservices.sourceforge.net`.

The primary purpose of this WBEM server is to simplify the management of the Solaris operating system, but it can, of course, be used to manage other parts of a system running on Solaris.

b4wbem

This project, which was started as a management interface for the Linux operating system, appears to have been moribund since 2001. See `http://b4wbem.sourceforge.net/` for details.

Tools

One important characteristic of the WBEM/CIM architecture is the standard CIM-XML interface between WBEM clients and WBEM servers. This means that you can pick-and-mix clients and servers: use, for example, a test client from one WBEM implementation to test a server from another. I regularly use the WBEM Services Java client to access the openPegasus C++ WBEM server.

Generic WBEM Clients

CimNavigator

CimNavigator is a WBEM client written in Java which provides a generic graphical user interface into a WBEM server, allowing the user to manipulate the schema, create and modify instances and name-spaces, and invoke extrinsic methods.

Being written in Java, it is very portable and can be downloaded without charge (although the source code is not made available) from http://www.cimnavigator.com.

Most of the tools described in this chapter have been written by companies but CimNavigator has been written and released by an individual, Andy Abendschein.

SBLIM CLI Client

The SBLIM project has released a generic CLI called wbemcli. This is a stand-alone program with an input and output syntax suited for use by Shell and Perl scripts and is available from the SBLIM home Web site: http://www-124.ibm.com/sblim/index.html.

SBLIM Event Tool

SBLIM has also released a tool, called evsub, which is a client application that automates the process of setting up subscriptions to indications. The package also provides indication handlers which signal the indications as they arrive by a creating a pop-up menu, sending an e-mail, executing a shell command or invoking a method on a particular instance of a class. This tool takes a lot of the labour out of testing indication providers.

openPegasus Browser Client

The openPegasus distribution includes a generic Web browser-based client which can be used for test purposes. At the time of writing it is effectively moribund and incomplete—most users of openPegasus using instead the WBEM Services client. The openPegasus client does, however, support the operations of retrieving and listing classes and instances.

openPegasus CLI Client

openPegasus also includes a very simple generic CLI called `cimop`. In principle, this allows you to issue any of the intrinsic commands. In practice, not all of these have currently been implemented: in particular, `enumerateClasses()`, `invokeMethod()`, `createClass()`, `modifyClass()`, `associators()`, `references()`, `getQualifier()`, `setQualifier()`, `execQuery()`, `referenceNames()`, `associatorNames()`, `deleteQualifier()`, and `enumerateQualifiers()` are not implemented.

WBEM Services Client

As part of their WBEM Services product, Sun releases a generic WBEM client, known as CIM Workshop which is a graphical application that allows you to browse the CIM schema and to perform operations on CIM classes and instances.

Code Generation Tools

SBLIM

As part of the SBLIM project, IBM has made available a program, known as `psg`, which generates skeleton code for providers in CMPI C, NPI C and NPI C++ formats.

As part of the same project, IBM has also released a plug-in for Rational Rose which allows CIM models to be imported to and exported from Rational Rose in *mof* format.

Vintela

Vintela Inc has released a Java-based code generation tool called `CodeGen` which uses the Velocity Template Engine (see `http://jakarta.apache.org/velocity/`). `CodeGen` surprisingly does not use the *mof* files to find the necessary details of the classes—instead it connects to a WBEM server into which the class has been loaded and extracts the details from it.

`CodeGen` can produce provider stub code or documentation in this manner. It is available from `http://sourceforge.net`.

Server Test Tools

Probably the most useful set of WBEM server tests are those being created by the SBLIM project. This project has developed a test suite including a test infrastructure which you can use to automate your regression testing. The test suite is well documented in a report which can be downloaded from `oss.software.ibm.com/sblim/doc/SBLIMTestSuite.pdf`. The test suite covers three types of test:

- Interface tests to test the providers: do they accept the complete set of parameters, do they correctly return exceptions when they cannot (or should not) carry out the operation?
- Consistency tests to test the overall consistency of the management view presented by the providers and WBEM server: is the number of instances of a particular class always the same irrespective of the way in which they are enumerated, are property values consistent with their specification?
- Specification tests which use the meta-information about the model to check whether providers implement the class definitions properly: if a property is marked as required, does the provider always return it?

The openPegasus implementation contains a number of simple but useful test tools. Since the openPegasus source is freely available, these tools make useful starting points for you to develop more sophisticated ones. In particular openPegasus comes with a useful scripting tool called `wbemexec`. This program acts as a WBEM client and takes a file containing XML (valid or not) and sends it to the WBEM server. Whatever is returned is displayed on *stdout*. This is really useful for checking the WBEM server's response to invalid XML and for checking the reaction of providers to invalid or edge-case parameters.

mof *Editors, Browsers, and Syntax Checkers*

CIM Compatibility Checker

Intel has made available a CIM Compatibility Checker known as CCX. This is now a little long in the tooth, but it still provides a platform-independent application for testing a CIM implementation against the DMTF's core and common models. Two tests are provided:

1. CIM Class Definitions testing
2. CIM Class Instance and Associations testing

In both tests, schema definitions and class definitions such as class properties and associations are verified for existence and correct data type.

CIMValidate

There is a number of CIM tools available on the home page of Langdale Consultants (`http://www.langdale.com.au/`) including CIMValidate, a tool for validating CIM power system models developed by the Electric Power Research Institute (EPRI).

This site provides an interesting viewpoint on WBEM/CIM as the introductory text on the home page says, "The CIM/XML language is a language for representing power system models." SNIA, of course, thinks that CIM is a language for representing storage area networks and in the telecommunications industry we see it as a language for representing network components and services. It is good to see these disparate views since they reflect the wide applicability of WBEM/CIM.

Although driven by Langdale Consultants, CIMValidate is an open source project to which anyone can contribute. The project is maintained on sourceforge.

WMI Software Developer Kit

As part of their support for WMI, Microsoft has developed the WMI Software Developer Kit (SDK). Unfortunately not only does this require a computer running Microsoft's Windows operating system to run, it needs a Microsoft Web browser even to download it!

The WMI SDK includes a program, CIM Studio, which can be used to browse CIM models.

Microsoft also makes available a simple *mof* editor to allow *mof* files to be produced somewhat more easily. This is not actually a graphical editor; instead it is a text-based editor specialised for *mof* programs. This tool is available from the DMTF Web site but it comes with quite an onerous licence including the term: "You may not rent, lease, or lend the software product" which would seem to imply that every member of a team must download his or her own copy.

WBEM Solutions SDK

WBEM Solutions Inc also produces and sells a WBEM SDK which includes automated code generation as well as various generic clients.

Solaris WBEM SDK

Sun Microsystems has released a software development toolkit (SDK) with its WBEM server.

mib2mof

Sun Microsystems has also released a tool, `mib2mof`, which converts Solaris SNMP MIBs to *mof* files.

mof2html

Nortel Networks has released a simple tool into the public domain called `mof2html`. This tool, which is written in C++, should compile on almost any operating system (it has been tested on various versions of Linux and SUN Solaris). If it is executed pointing to the top-level *mof* file of a DMTF release (normally called CIM_SchemaXX.mof where XX is the version number, e.g., CIM_Schema28.mof) then it creates html files allowing the various classes and properties to be browsed with a Web browser with hot links to superclasses, etc.

Chapter 16

Choosing WBEM Software

If you are about to embark on a WBEM/CIM project within your company then you will obviously need to select the appropriate WBEM server software. This raises the normal "build or buy" debate and it is perhaps worth looking at the effort involved in building and supporting the basic WBEM software.

Home Brew

As the standards are firm you may be tempted to build your own WBEM server and associated tools. For your particular products this may be the best way to go but, when estimating the size of the development, bear the following numbers in mind.

The openPegasus release 2.2 code consists of about 114,500 non-comment lines of C++ broken down as shown in Table 16.1. Although this represents only one implementation, I have no reason to believe that it is atypical.

If the COCOMO II* model for estimating the time required to develop software is retrospectively applied to this, and if appropriate values are taken for the productivity factor (2.58) and the penalty (1.11), then this represents an effort of about 40 programmer years.

Continuing with the COCOMO II model, the effort required to maintain this amount of code can be estimated at 14 programmers

---

* See http://www.softstarsystems.com/cocomo2.htm.

Module	Approx Lines
Server	102,000
mof Compiler	6,500
Client Libraries	6,000
Total	114,500

Table 16.1   openPegasus v2.2 Lines of Code

in the first year, dropping to a steady-state maintenance team of 7 programmers after 4 years.

These are fairly large numbers for any company and, if you are thinking of developing your own WBEM system, you need to be willing to spend this amount of money. On the other hand, if you buy commercially, then you need to ensure that the supplier has dedicated a team of approximately this size to the ongoing maintenance of the product. Remember that these numbers are for the WBEM server only; they do not include any allowance for writing your own application—the clients, providers and listeners.

Reviewing a Bought-In Product

If the above deters you from developing your own CIM server then you will need to obtain one from the open source community or from commercial vendors. Questions you might like to ask about potential sources include the following:

- What WBEM operations does the implementation provide? In particular, are the more recent features such as queries supported?
- In which language is the code written? Does it lend itself to an embedded environment (if this is important to you)?
- How does the implementation interface to providers? Is the CMPI supported? Is each provider executed in a different thread? Is a provider executed in an address space different from that of the WBEM server so that an error in the provider cannot cause the server to crash? Are providers dynamically linked as required? If so, are they automatically unlinked after a given time? If so, then can they be pinned in memory to reduce the overhead of having to save state?

- How well has the implementation been adopted in the industry? A widely adopted implementation is likely to be better supported than one with little uptake. Is there a regular release schedule for the implementation and is the feature roadmap published in advance?
- Is the code open source? If not, what guarantees does the supplier offer for continued support? If so, then what are the licencing conditions and are these acceptable to your company? If the product is not open source then does the vendor make the source available to you? If not, how many different processor/operating system/board combinations will the vendor support?
- How easy is it to install? Does the package contain test cases which can be run to ensure that the installation has been successful?
- What administration and developer tools are available?
- What levels of security are provided with the implementation?
- How much memory does the implementation require (particularly important if the code is to be implemented in an embedded device)?
- What performance benchmarks are available for the implementation? A poorly designed repository interface, for example, can seriously impact the performance of an otherwise good implementation.

Open Source

As for all such projects, the open source WBEM/CIM projects listed in Chapter 15 come with a varying level of support: ranging from the excellent to the nonexistent. Posting a question to a mailing list will allow you to gauge the activity within a project and I would advise you to monitor a group's mailing list and attend teleconference meetings before committing yourself to a particular implementation.

Another question to consider when using open source software is how much influence your company is likely to have over the direction of the product. If you buy a commercial product then presumably you can offer additional money to the vendor to encourage the introduction of the new features you require. In the open source world, the features that you find essential may not be of interest to others—are you prepared to design and code them yourself and then release the results to the wider community?

Note that a project being open source does not mean that it is uncontrolled. The features of openPegasus, for example, are tightly controlled by a steering committee and code is released at regular intervals with the content of each release being defined and agreed in advance (in Project Enhancement Proposals (PEPS) which can be viewed on the openPegasus Web site). In addition, one of the companies represented on the steering committee provides a project manager to control the content, quality and schedule of each release. Although the results are open source, the control of the development process is probably as tight as on any other software project.

Commercial

There are numerous companies offering different levels of commercial support ranging from a full WBEM server implementation to limited consultancy. I have just used Google to search for the keywords "wbem consultancy" and had 280 hits, most of which, according to my unscientific sampling, appear to be consultancy companies offering support for WBEM.

Whether or not commercial support is required for your project is, of course, a decision that only you can make but here are a few points which you may wish to consider:

- As you can see from Chapter 9, creating your first CIM model is not trivial. Determining precisely what your classes are and where they fit in the DMTF's hierarchy is not trivial. Consultants, familiar not only with the results of the standardisation process (the standards) but also with the discussion which went on to reach those results, can be invaluable.
- Many commercial companies offer the service of taking the open source software products and charging for a more controlled release cycle, thorough regression testing, emergency bug fixing and better documentation. This may be a suitable half-way house between true open source and commercial products.
- Some of the commercial WBEM servers have been developed for a specific market which might match your requirements better than a general-purpose WBEM server. There is, for example, at least one C-based WBEM server on the market which has a much smaller memory footprint than any C++ or Java implementation.

APPENDICES

APPENDICES

Appendix A

Industry Adoption

WBEM/CIM is currently a hot topic in a number of areas, including:

- **The Storage Network Industry,** which has adopted the standards enthusiastically—see, for example, the Storage Networking Industry Association (SNIA) Web site (`http://www.snia.org/home`). Roger Reich, a Director of SNIA, has said, "The SNIA has set a goal that all new storage networking products will utilize the SMI-S standard in 2005. Our recent announcements demonstrate that the SNIA is accelerating down the path to making this goal a reality."

    SMI-S, which used to be known as Bluefin, is an implementation of the WBEM/CIM architecture.
- **The Global Grid Forum,** concerned with the establishment of standards for Grid Computing, has defined CIM models for its services: see their Web site (`http://www.globalgridforum.org/1_GIS/CIM.htm`).
- **The Electrical Power Supply Industry** (a different sort of grid computing?). The Electric Power Research Institute (EPRI), an international body with the mission of "providing science and technology-based solutions of indispensable value to its global energy customers by managing a far-reaching programme of scientific research, technology development, and product implementation," has developed a CIM-based model of electrical power distribution. This has been implemented by a number of electrical power utilities and is in the process of being adopted by the International Electrotechnical Commission (IEC) as an energy management system standard (IEC 61970-301).

■ **The Telecommunications Industry**, which appears to be adopting CIM models for services and devices to facilitate better interworking and conformity at the management interface. Naturally, much of this is arising in the products where storage area networks meet telecommunications: for example Cisco Systems' MDS9000 product family. It is possibly a leaking of the WBEM technology from the storage area network into the associated telecommunications network which will drive WBEM deeper into telecommunications.

There is another piece of evidence supporting the view that a move is being made within the telecommunications industry towards WBEM: the announcement of formal ties between the DMTF and two bodies from the telecommunications industry, the TeleManagement Forum (TMF) and the Service Availability Forum (SAF).

The TeleManagement Forum is a nonprofit group of over 300 companies related to the telecommunications industry including service providers, computing and network equipment suppliers, software solution suppliers, and their customers. It is dedicated to providing architectures to improve the management of information and communications services (see `http://www.tmforum.org/`). The forum has devised an architecture, known as New Generation Operations Systems and Software (NGOSS), which is an integrated framework for developing, procuring and deploying [management] systems and software. The following was contained in a joint TMF/DMTF press release in June 2003.

> "The TM Forum and DMTF have traditionally been major players in the telecommunications and enterprise industries, respectively, and both have recognized the industry trends and drivers toward convergence [...] They are now collectively working on the potential for integration of these shared information models to also support the essential need for telecommunications/enterprise convergence.

> Examples of converged services include voice over IP (VoIP) and the integration of other forms of traditional connection-based and connectionless services. These new generation services are increasing the need to manage computing and storage resources in the telecommunications environment, and to operate telecommunications and enterprise hardware and software in an integrated fashion," said Winston Bumpus of Novell, president, DMTF. [...]

> "The line between the enterprise and telecom is becoming irrelevant as customers seek to have end-to-end management

of devices, services and applications," said Martin Creaner, vice president and CTO, TM Forum. "Although TM Forum and DMTF approach the development environment differently [...] both organizations are working on the same basic areas— that of systems, physical entities, software and applications, services, statistics and policies. Both organizations also have embraced an object-oriented design approach, with extensibility by both the organization and vendors as a key goal. [...]" Whether this DMTF/TMF marriage has resulted from genuine affection or is a marriage of convenience remains to be seen.

The SAF (see `http://www.saforum.org/home`) is a also a coalition of communications and computing companies working to create and promote open interface specifications. Its *raison d'être* lies in the traditional difference between "Enterprise" grade systems and the more highly-available "Carrier" grade systems. Increasingly, techniques common in the enterprise but previously unknown in the carrier market are beginning to impinge on carriers. In particular, the transition to packet-based, converged, multi-service networks means that carrier-grade infrastrucure techniques need to be applied to these networks. This requires interoperable hardware and software building blocks, management middleware, and applications implemented with standard interfaces.

It appears that the SAF has decided to embrace the WBEM standards as part of its management strategy. This may be the necessary opening for WBEM to find its way out of the enterprise and into the carrier's world.

- **The Desktop Computing Industry.** This is where CIM started life and the predominant desktop operating systems, Linux, Solaris, HP-UX, and Microsoft Windows, all have CIM models, WBEM interfaces and providers. Various commercial products, such as LANDesk, previously supported by Intel and now owned by LANDesk Software, Inc. are available to manage desktop computers containing WBEM servers.

Microsoft uses the term "Windows Management Instrumentation" (WMI) for its implementation of WBEM. Unfortunately, although Microsoft's implementation does conform to the DMTF schema definitions, it does not interoperate with other CIM technologies because it uses Microsoft's COM/DCOM interface rather than the standard CIM-XML. Mappers do, however, exist to permit interworking. In particular, Vintela has released a product known as VMX (effectively a provider) which allows the

Microsoft Systems Management Server (SMS) to manage UNIX and Linux systems.

IBM has released its "Standards Based Linux Instrumentation for Manageability" (SBLIM: pronounced "sublime") which is "intended to enhance the manageability of GNU/Linux systems." The goal, apparently achieved, is to enable the practical management of Linux systems using WBEM technologies.

Sun Microsystems has also announced its adoption of WBEM and its accompanying standards (CIM, XML, etc.) for Solaris and has said that "WBEM is part of Sun's long-term strategy for system management." This is reflected in its release of a WBEM Software Development Kit (SDK).

Appendix B

"Is-A" and "Has-A" Relationships

I claimed on page 56 that one of the important differences between CIM and systems such as SNMP is the use of *is a* rather than *has a* in the modelling process. This appendix tries to justify that claim by considering the difference between these two ways of modelling: by comparing the relationship that "A is a B" (a shelf is a System Component) with the relationship "A has a B" (a shelf has cards).

To illustrate the difference between the approaches, we'll return to Joe's dog ownership which we discussed in Chapter 5. For your convenience, I have reproduced Figure 5.3 as Figure B.2.

Figure B.1 shows part of a diagram based on *has a*. It shows, for example, that a Human *has a* (or may have a) Domestic Dog. Note that this does not indicate that any particular human has a dog, only that dog ownership is something that may be associated with a human.

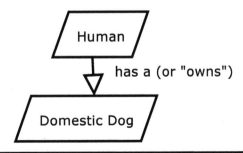

Figure B.1   A Simple Has-A Relationship

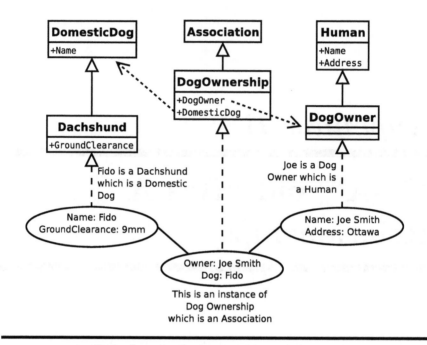

Figure B.2   Fido with an Association (Is-A)

In a more complex example, the information shown in Figure B.1 might be enough to allow an operator to associate a particular human with a particular dog, whereas an operator trying to associate a particular human with a particular elephant could be rejected with an error message because it does not fit the model.

Note that Figure B.1 could equally well be inverted to indicate that a Domestic Dog *has an* owner.

Contrast Figure B.1 with Figure 5.2 on page 52. In the case of Figure B.1, the basic relationship between the classes is one of ownership (*has a*) whereas in Figure 5.2 it is one of *is a*.

Of course we need to add the actual instances of Joe and Fido to Figure B.1; see Figure B.3. You might think that it is unnecessary to create an association between Joe and Fido since Humans are able to own Domestic Dogs, Joe is a Human and Fido is a Domestic Dog; therefore Joe owns Fido. This breaks down, however, when we add Leslie. She owns a Domestic Dog called Spot. When we add Spot and Leslie to the diagram (see Figure B.4), it is no longer clear who owns which dog. To indicate this, we add *ad hoc* associations as in Figure B.5.

Now I will complicate matters further by assuming that there is some sort of relationship between Joe and Leslie; perhaps Joe joined

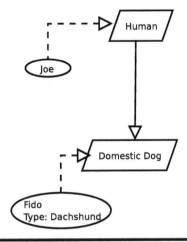

Figure B.3 Adding Instances to Has-A

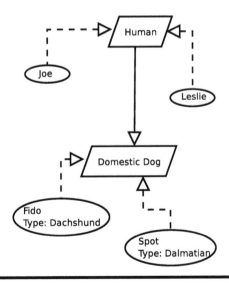

Figure B.4 Connecting the Instances with Has-A

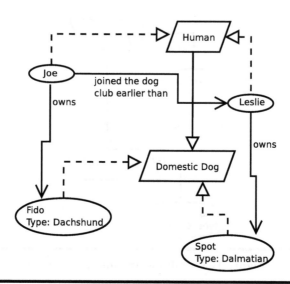

Figure B.5   Adding Another Association with Has-A

the local dog club earlier than Leslie, and we need to model it. We now have to add another *ad hoc* relationship as also shown in Figure B.5.

It is now unclear precisely what purpose the original lines joining the classes in Figure B.1 are serving. If they were removed from Figure B.5, then would anything be lost from the diagram? The answer is that, in the manner that *has a* diagrams are used in device management, one important thing *would* be lost: addressability. There may be several items called Spot in a larger diagram and to identify the Domestic Dog called Spot uniquely I could use a name such as Human.DomesticDog.Spot. If the lines joining the classes together are removed then some other way must be found for identifying a particular instance uniquely.

The *has a* class diagram effectively gives prominence to one association between humans and domestic dogs, the *has a* (or "owns") relationship. Other relationships, for example, *is heavier than* or *is married to* or *joined the dog club earlier than*, have to be handled in an *ad hoc* fashion.

The *has a* representation also has another drawback: as can be seen from Figure B.5, it effectively requires one tree for classes and a second tree for instances. For clarity, I have separated them in Figure B.6.

The *has a* relationship is the conventional (SNMP) way to represent management information: a rack *has a* shelf which *has a* slot which *has a* card which *has a* processor which *has an* instance of the OSPF routing software which *has a* routing table, etc.

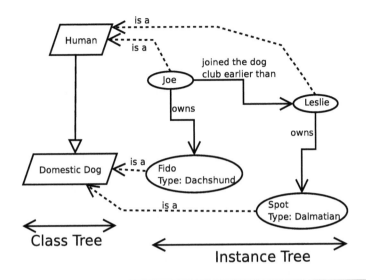

Figure B.6   The Class and Instance Trees

As we have seen, however, this approach leads to two separate tables: one containing the class structure ("a card can have a processor but a processor cannot have a rack") and one containing the particular instances of the classes. Here, the primary relationship is really *is on*.

This technique was applicable to the management of hardware because the items being managed were effectively static—a card *is on* a shelf is probably inviolable—but is less applicable to software, which is mobile, and services, which are much more abstract. It is, of course, necessary to express the card/slot/shelf *has a* relationship in CIM and this is handled by using associations; see Figure B.7.

In summary, several major distinctions between Figure B.2 and Figure B.6 can be drawn:

- In the *is a* model, there is one tree containing both the class definitions and the instances.
- If another dog and owner were to be added to the system, then this would effectively form a new instance sub-tree in Figure B.6—the instances could be drawn in a self-contained manner with *owns* relationships. If a new dog and owner were to be added to Figure B.2 then the owner and dog would be distributed throughout the diagram. This makes the additions harder but the final result more integrated.
- Instances are handled more cleanly in the *is a* model of Figure B.2. Instances are clearly objects derived from a particular class;

the separate tables for classes and instances are not required. In the *has a* model of Figure B.6 their precise status is unclear: they effectively form a separate sub-tree.

■ In the *has a* model, the manipulation of classes ("add a new class of routing algorithms") is basically different from the manipulation of instances ("add another OSPF instance to processor 23")—the former having to be done by a programmer and the latter by an operator. As system management moves towards service management, the class structure will need to be modified by the operator rather than by the equipment supplier and the combination of the class and instance tables will become a useful simplification.

■ Particular relationships (e.g., marriages) are instances of first-class objects in Figure B.2 and can therefore be manipulated as objects ("list all Spot's relationships" or "list all people who have a domestic dog").

The addition of a relationship (e.g., "is married to") is quite difficult and somewhat *ad hoc* in the *has a* model; each instance needing to be changed to provide another pointer to the specific person (i.e., instance) to whom he or she is married. It is not clear where the rules about this relationship (men/women may only marry women/men,* a man/woman may only be married to one woman/man, a man/woman need not be married to any woman/man) are to be held. If an attempt is made to add the relationship "A is married to B" then this needs to be rejected if A is already married to C.

In the *is a* model of Figure B.2, by contrast, the *is married to* relationship could clearly contain the rules regarding the creation of the relationship, for example, that an *is married to* relationship cannot be created for a person if that person already has one; that would be bigamy! These rules would be inherited by all instances (and A's proposed illegal marriage to B would be detected and prevented).

These differences between the two representations are summarised in Table B.1.

---

* Except in Canada.

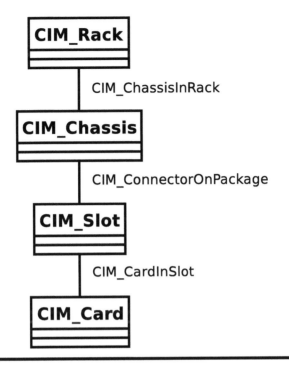

Figure B.7   Has-A in CIM

	Has A	**Is A**
Abstraction	Structural	Behavioural
Inheritance	Subtyping	Subclassing
Number of Trees	Two: instances handled separately from classes	One: instances handled with classes
Adding new type of device	Easy: make new sub-tree	More difficult: determine where the new classes fit into existing model
Management of	Devices	Devices, processes and services
Relationships	*ad hoc* and static	Integral and malleable

Table B.1   Comparison of *is a* and *has a*

Appendix C

FCAPS

I briefly introduced the so-called FCAPS functions in the introduction to this book. I give a little more detail here about this classification of device management.

FCAPS arises from the ITU-T's TMN layering structure and comprises the initial letters of fault, configuration, accounting, performance, and security management:

- **Fault management:**
  - The detection that a fault has occurred.
  - The filtering and correlation of the fault reports. In many systems the number of fault reports arising from a single failure can be enormous: if these are not filtered then both the reporting system and the operator can be overwhelmed. Good filtering and correlation allow management systems to grow without becoming unwieldy.
  - The signalling of root cause to an operator.
  - The opening of trouble tickets and the management of these (particularly when the fault has occurred in hardware) until the defective component is replaced. However, because most faults occur in software, the replacement of faulty parts is normally not an issue; the failed module can be reloaded and restarted.
- **Configuration management:** allowing the operator to make (or suggest) configuration changes for the device: bring this port up, take that port down, change the IP address associated with this other port. Configuration management also includes the

auto-discovery of components and capabilities by the system itself.

- **Accounting management:** collecting and safely storing the information needed to bill customers. This is naturally of prime importance to the device owner and may be surprisingly difficult to provide in many applications—records may have to be held securely on a device for a long time if contact is lost with the central accounting system.
- **Performance management:** collecting performance information (e.g., number of packets passed per second from a particular port, maximum and mean lengths of packet queues, number of incoming requests handled per second, etc.), compiling and consolidating this into a form that allows a human to identify issues affecting the present and potential performance of the managed devices.
- **Security management:** controlling access to the managed device by users (the people using the device) and operators (the people managing it). Security management includes authenticating users and operators to ensure that they are who they say they are and checking their authority: the level of access they are allowed to the system. Security management may also be involved in protecting the device from malicious denial-of-service attacks.

Another acronym which is sometimes used in conjunction with FCAPS is FAB, fulfillment, assurance, and billing:

- Fulfillment (a strange word) covers planning; anticipating the need for additional or reduced capacity and ensuring that it is installed or removed as required.
- Assurance covers what would be considered performance, configuration, and fault management in the FCAPS model; i.e., the tasks needed to ensure that customers are getting the services they are paying for.
- Billing which covers the collection of payment from the customers for the services they are receiving; i.e., accounting in the FCAPS model.

Appendix D

Miscellaneous Datatypes

The datatypes used within *mof* are relatively straightforward, with the exception of `datetime` and `string`. This appendix describes these.

The `datetime` Datatype

This datatype can be used to represent either a specific time or an interval between specific times. Both of these representations consist of a string of exactly 25 characters. A specific time is expressed in the form:

```
yyyymmddhhnnss.uuuuuuPxxx
```

and an interval is expressed in the form:

```
ddddddddhhnnss.uuuuuu:000
```

where the fields are as shown in Table D.1. All fields must be zero-padded on the left to preserve their length and unused fields must be replaced with the appropriate number of asterisks.

For example, 5 minutes, 23.243424 seconds after midday on my birthday, 14th November 2004, in Ottawa (EST), would be represented as

```
20041114120523.243424-300
```

Field	Contents
yyyy	4-digit year
mm	2-digit month (01 to 12)
dd	2-digit day (01 to 31)
hh	2-digit hour (00 to 23): 24 hour clock
nn	2-digit minute (00 to 59)
ss	2-digit second (00 to 59)
uuuuuu	6-digit representation of microseconds
Pxxx	is a representation of the number of hours between the local time and UTC (Universal Co-ordinated Time) in the format ±xxx where xxx is a number of minutes (points west of Greenwich being negative, those east positive). This field is always set to ':000' in the representation of a time interval: it is the colon which distinguishes an interval from a specific time.
dddddddd	is a number of days

Table D.1   The "datetime" Fields

Similarly, an interval of 10 days, 17 hours, 4 minutes, and 17.0887 seconds would be represented as

```
00000010170417.088700:000
```

The string Datatype

The CIM **string** datatype consists of a number of UCS-2 characters.

UCS-2 is based on the Basic Multilingual Plane (BMP) of ISO/IEC 10646 and is basically a 16-bit encoding of a large number of Unicode characters—international alphabetic characters and symbols. UCS-2 includes the well-known ASCII characters as a sub-set, mapping them to their "normal" ASCII equivalent values, but also contains a large number of European and Asiatic characters.

Thus a capital 'A' is encoded as 0x0041 as it would be in ASCII. The Greek letter β, on the other hand, does not appear in the ASCII character set but does in UCS-2—as the encoding 0x03B2.

Strings within CIM are composed of UCS-2 characters rather than ASCII characters.

More information about UCS-2 and Unicode can be found on the Web site of the Unicode Consortium (http://www.unicode.org/); see also the glossary entry in this book for Unicode.

Appendix E

The MappingStrings Qualifier

`MappingStrings` is a qualifier which may, in principle, be attached to any CIM declaration, but which, in practice, is normally used to qualify a property. It is used to define an equivalence (a "mapping") between the property to which it is attached and a related property in another standard.

For example, as you would expect, the International Telecommunication Union (ITU-T) has a number of standards related to "network pipes," for example, optical connections between devices. The DMTF description of a pipe contains the *mof* shown in Figure E.1 on the following page.

This figure illustrates the MappingString qualifier on both a class (CIM_NetworkPipe) and a property (Directionality). In each case the mapping string itself consists of the name of the standards body (ITU in this case) followed by a vertical bar and then the definition of the property within the other standard. Although not shown in Figure E.1, a further vertical bar can be added to the string and then the version number of the other standard.

As Figure E.1 also shows, there may be more than one mapping string: perhaps a property appears in both an ITU and an IETF specification, in which case both strings would be present.

```
[Version ("2.7.0"), Description (
    "NetworkPipe is a subclass of EnabledLogicalElement, "
    "representing the state and management of a "
    "connection or trail between endpoints.  " ),
 MappingStrings {"Recommendation.ITU|M3100.Pipe",
    "Recommendation.ITU|M3100.TrailR1",
    "Recommendation.ITU|M3100.ConnectionR1",
    "Recommendation.ITU|M3100.SubNetworkConnection"} ]
class CIM_NetworkPipe : CIM_EnabledLogicalElement {
    [Description (
        "Indicates whether the pipe is bi-directional "
        "(value = 2), unidirectional (value = 3), or this "
        "information is not known (value = 0).  For "
        "unidirectional pipes, the source and sink are "
        "indicated by a property (SourceOrSink) of the "
        "association, EndpointOfNetworkPipe."),
    ValueMap {"0", "2", "3"},
    Values {"Unknown", "Bi-Directional", "Unidirectional"},
    MappingStrings
        { "Recommendation.ITU|M3100.Pipe.directionality"} ]
    uint16 Directionality;
```

Figure E.1   Example Usage of MappingStrings

Frequently Asked Questions

FAQ 42 *I have two classes, classA and classB. classB is a subclass of classA. classA contains a mappingString and so does classB. Both refer to the same IETF specification. What implication does this have for the two mapped entities—how do they need to be related in the IETF specification?*

They do not have to be related. This mapping is not a structural equivalence. It is simply a property-by-property mapping.

Appendix F

Installing openPegasus

The example given in Chapter 12 makes use of the openPegasus WBEM server. Obtaining and installing this software is very easy and, to allow you to reproduce the example, I give a blow-by-blow account here. I have borrowed heavily in this appendix from a document, "Writing a Pegasus CIM Provider," that was released by Nortel Networks into the public domain in October 2002.

openPegasus can be compiled for a number of different platforms, including Linux, openVMS, HPUX, Z/OS, OS/400, AIX, Windows XP and Windows 2000. The process I describe in this appendix is more applicable to the UNIX-like operating systems on this list, particularly to Linux, but the general principles should hold for all.

openPegasus relies on the presence of remarkably few other software packages but it does require **gnumake** to be installed. This is available from http://www.gnu.org. In its turn, **gnumake** relies on a few UNIX utilities that are not indigenous to computers running the Microsoft Windows operating system. The openPegasus team has therefore built a utility program, MU.EXE, to provide equivalent utilities for that environment.

Obtaining openPegasus

Pegasus can be checked out of the CVS system on the Pegasus Web site (http://www.openpegasus.org). Anonymous access for read is possible with username and password both set to **anon**:

```
CVSROOT=:pserver:anon@cvs.opengroup.org:/cvs/MSB
```

The source tree is in the directory **pegasus**. To check out the complete Pegasus source tree, type:

```
cvs co pegasus
```

A Pegasus directory will be created under the current directory and populated with the complete source tree and documentation.

To get the latest updates after a checkout, just type the following from the directory in which you issued the checkout command:

```
cvs update -d
```

Setting Environment Variables

The installation and execution of openPegasus requires certain variables to be set in the user's environment. These are as follows, assuming that $HOME is set to point to the directory into which openPegasus has been downloaded.

```
export PEGASUS_HOME=$HOME/pegasus
export PEGASUS_ROOT=$HOME/pegasus
export PEGASUS_PLATFORM=LINUX_IX86_GNU
export PATH=$PATH:$PEGASUS_HOME/bin
export LD_LIBRARY_PATH=$LD_LIBRARY_PATH:$PEGASUS_HOME/lib
```

$PEGASUS_HOME is set to point to the parent directory for the generated bin, lib, etc. directories.

$PEGASUS_ROOT is set to point to the parent directory for the source code (as checked out of the openPegasus CVS). I have made $PEGASUS_ROOT and $PEGASUS_HOME the same in this example but this is not necessary.

$PEGASUS_PLATFORM is set to a value to define the type of platform being used. The example I give above is for Linux running on an Intel processor. There are many other options including WIN32_IX86_MSVC, AIX_RS_IBMCXX, HPUX_PARISC_ACC and ZOS_ZSERIES_IBM.

Compiling openPegasus

Before trying to compile openPegasus, ensure that the environment variables have been set. Then change directory to **$PEGASUS_ROOT** and type **make**. This (eventually—go and make a cup of coffee) builds the WBEM server, *mof* compiler and other utilities.

Loading the Repository

Before making use of the system, it is useful to have the core and common models loaded into the repository. To do this, go into the directory **$PEGASUS_ROOT/Schemas** and type:

```
make repository
```

This command can be used later to remove and reload the repository. To add the necessary linkages in the InterOp namespace, go into the **$PEGASUS_ROOT/Schemas/Pegasus** directory and again type:

```
make repository
```

Loading an Example Application

openPegasus comes with a number of example providers; for simplicity I will demonstrate how to load those for the Linux operating system. To build the Linux providers go into the directory **$PEGASUS_ROOT/src/Providers/linux** and again type **make**.

Using these providers will require that the Linux classes be in the repository. To achieve this, go into the directory **$PEGASUS_ROOT/-src/Providers/linux/load** and type:

```
make repository
make registration
```

Running the WBEM Server

Once all this compilation and loading is complete, the WBEM server, **cimserver**, is ready to be started. Type:

`cimserver [ [ options ] | [ configProperty=value, ... ] ]`

where the options include:

- -v to display the CIM server version number
- -h to print a list of options
- -s to shut down the CIM server (you must be logged in as root to enter this command—otherwise use the kill command)
- -D [home] to set Pegasus home directory (which is not needed if the environment variables given above are set)
- configProperty=value to set a CIM server configuration property (for a list of the keywords which may be used here, run `cimconfig -1`)

Appendix G

Glossary

This glossary is intended to elucidate acronyms and concepts that may not be completely understood by all readers. It does not include terms explained elsewhere in the book; it does include those terms with which I assume that the reader is familiar. To learn more about a specific topic which *is* covered in the main text, the index may be a better starting point.

BGP. Read the description of OSPF in this glossary and then return here. BGP—Border Gateway Protocol—is also a protocol by which routers exchange information about hosts they can reach. Unlike OSPF, however, it is normally used between routers belonging to different administrations.

CGI. Common Gateway Interface. You can think of a CGI program (sometimes called a script) as a dynamically created Web page. When I direct my browser to a particular URL, http://www.slashdot.org for example, my request arrives at a Web server which has to decide what to do with it. Often there is a preformatted HTML page available and the Web server simply picks this up and returns it to the browser for display. Otherwise, it is possible to direct the Web server, not to an HTML page but to a program, a CGI program. This program is then responsible for creating HTML and returning it to the Web server which sends it to the browser for display. In the WBEM/CIM context, Figure 13.3 on page 255 illustrates a Web browser talking to a Web server which invokes the WBEM client which is acting as a CGI program. The WBEM client is then responsible for communicating with the WBEM server and returning an HTML page for display.

CGI is therefore not a computer language—CGI programs can be written in C, C++, Python, Perl, Fortran, or any other language able to manipulate data and return an HTML "page."

CIM. Common Information Model. This term covers the language and methodology used to define an implementation-independent management models (*mof*) in the context of the core and common models developed and published by the DMTF.

CLI. Command Line Interface. I use this term to describe an operator interface which involves typing a command into some form of terminal or computer and getting textual information back. This contrasts, for example, with a Graphical User Interface (GUI) where the same command might be entered by selecting it from a pull-down menu and the response displayed graphically.

CORBA. CORBA is a confusing term to use in a book about interfaces between management clients and servers since it is itself a management interface. I do not use it in that way in this book, rather as a sophisticated form of remote procedure call. CORBA stands for "Common Object Request Broker Architecture" and is standardised by the Object Management Group (OMG). By introducing an Object Request Broker (ORB) into a system a client is able to invoke a possibly remote server transparently, irrespective of the programming languages, processor types (Endianness) or operating systems of the client and server.

In much the same way that a WBEM server allows a WBEM client to invoke an extrinsic function on a provider, so does the ORB allow a general client to invoke a general server.

DMTF. Distributed Management Task Force (previously *Desktop Management Task Force*). This is the organisation which is responsible for the WBEM and CIM standards. These standards include:

■ The protocols operating between clients and servers, between servers and providers and between servers and listeners.

■ The core and common models.

The DMTF's Web site is at `http://www.dmtf.org`.*

IETF. Internet Engineering Task Force. This is the body which co-ordinates the work of a community of network designers, operators, vendors, and researchers concerned with the evolution

---

* Telecommunications engineers will find typing *dtmf* almost irresistable.

of the Internet architecture and the smooth operation of the Internet. It is open to any interested individual. The standards and discussion papers which emerge from the IETF are known as Requests for Comment (normally abbreviated to RFCs rather than the more correct RsFC).

mof. Managed Object Format—the formal language in which CIM models are written. I describe this language starting on page 64.

MPLS. When packets are sent across a network using the Internet Protocol (IP) there is no specific route that they will follow—packets flowing from the same source to the same destination may follow different routes. This is sometimes awkward because it does not allow the network to be correctly dimensioned for the anticipated traffic, some links possibly lying idle while others are overloaded. MPLS (Multi-Protocol Label Switching) is a protocol which allows "tunnels" with particular characteristics (bandwidth, delay, etc.) to be created through a network. Packets are then sent through the tunnels and all packets comprising one flow follow the same route, on which resources have been reserved.

OSPF. Routers in an IP network know about the devices (hosts) connected directly to themselves but initially have no knowledge about hosts elsewhere in the network. When a router detects a neighbouring router, it sends its local knowledge to it, receiving the other router's local knowledge in return. This allows each router to augment its table of known hosts and send a copy of the augmented table to its other neighbours. In this manner, the location of the various hosts in the network spreads like a plague throughout the network. Between the routers owned by one operator, the protocol typically used to exchange this information is OSPF—open shortest path first.

Opaque property. An opaque property is one with no structure that can be relied upon. The property (typically a string) must be treated *as is* and the user must make no attempt to understand its structure. Within CIM, there is a move towards opaque keys: see page 195 regarding CreationClassName and InstanceId.

RFCxxxx. Many of the standards quoted in this book have numbers like RFC3060. These are standards from the Internet Engineering Task Force (IETF) and RFC is the acronym for "Request for Comment." This is a somewhat misleading name but it has a long and honourable history. RFCs can be obtained from a number of sources, the easiest access point probably being `http://www.faqs.org/rfcs/`.

Semantic and syntactical information. At various points in this book I contrast semantic and syntactical knowledge, particularly as it relates to a model expressed in CIM.

"Syntax" is easy to understand, for example: *property examMark is an integer between 0 and 100 inclusive.* This can be expressed easily in most formal programming-style languages, including *mof.*

When you are asked for Joe's mark in the latest examination, the fact that it is property `examMark` in Joe's instance of the Student class that you need to access would be an example of semantic knowledge. It relates not to the format of the property but to its meaning.

For more information about this topic as it relates to the WBEM client, see page 248.

SNMP. The Simple Network Management Protocol is the dominant management protocol for data devices (IP routers, switches, etc.) within enterprises. I give a very brief introduction to its history and a comparison with WBEM/CIM starting on page 22.

Unicode. Unicode has its own home page on the Web, `http://www.unicode.org`, and the following description is extracted from that site.

Unicode is a universal character-encoding standard for all of the characters used in written, natural languages. The Unicode standard (and ISO/IEC 10646) supports three encoding forms that use a common repertoire of characters but which allow for the encoding of a million characters. This is sufficient for all known character-encoding requirements, including full coverage of all historic scripts of the world.

Currently the Unicode Standard defines codes for characters used today in the major written languages. Scripts include the European alphabetic scripts, Middle Eastern right-to-left scripts, and many scripts of Asia. It also includes punctuation marks, diacritics, mathematical symbols, technical symbols, arrows, dingbats, etc.

Altogether, version 3.2 of the Unicode Standard provides codes for 95,221 characters from the world's alphabets, ideograph sets and symbol collections.

XML. The letters stand for "Extensible Mark-up Language" and describe a textual (ASCII) language which can specify content independent of its form. Such mark-up languages have existed for a long time; this book, for example, is being written in LaTeX which is a markup language dating back 20 years or more.

In the days of print, documents appeared in one form—the form in which they were printed. Today, documents appear in many forms: as Web pages, as printed pages, as projected images, as e-mail content, etc. The same form may not be appropriate for all of these. A chapter heading, for example, may want to be centred in blue and displayed in a 24-point font on a projected slide, whereas in a book it may be more appropriate to have it left-justified in an 18-point black font.

The author should not need to be aware of all of these conventions; he or she should simply be able to specify that a particular piece of text is a title and leave the interpretation to the software using it. XML (and LaTeX) achieve this. In XML it should be sufficient to write:

```
<CHAPTERHEADING>
The Client/Server Interface
</CHAPTERHEADING>
```

and expect the publisher's software to present this in a manner which fits the house style.

If you look at a larger piece of XML, Figure 7.3 on page 128 for example, you will see the same basic structure. There is a message contained between MESSAGE tags:

```
<MESSAGE .....>
  ....contents of the message
</MESSAGE>
```

Within the message there are subelements, for example an IPARAMVALUE contained between *its* tags.

Because of its ASCII text nature and its tag structure, XML is not a compact language (compare, for example, the *mof* of Figure 12.1 and the equivalent XML of Figure 12.2) but it is extensible (you can add any tags with which you and the recipient of your message agree) and can be easily parsed by a computer program.

Appendix H

Licencing

DMTF Models

The actual *mof* code from the DMTF is subject to the following licence agreement:

> Copyright © 1998-2002 Distributed Management Task Force, Inc. (DMTF). All rights reserved.
> DMTF is a not-for-profit association of industry members dedicated to promoting enterprise and systems management and interoperability. DMTF specifications and documents may be reproduced for uses consistent with this purpose by members and non-members, provided that correct attribution is given. As DMTF specifications may be revised from time to time, the particular version and release date should always be noted.

openPegasus Code

Some of the code examples in this book, in particular the Widget example starting on page 211, make use of code modified from the open-Pegasus 2.2 release. This code is subject to the following licence:

> Copyright © 2000, 2001, 2002, BMC Software, Hewlett-Packard Company, IBM, The Open Group, Tivoli Systems.
> Permission is hereby granted, free of charge, to any person obtaining a copy of this software and associated doc-

umentation files (the "Software"), to deal in the Software without restriction, including without limitation the rights to use, copy, modify, merge, publish, distribute, sublicense, and/or sell copies of the Software, and to permit persons to whom the Software is furnished to do so, subject to the following conditions:

The above copyright notice and this permission notice shall be included in all copies or substantial portions of the software. The software is provided "AS IS," without warranty of any kind, express or implied, including but not limited to the warranties of merchantability, fitness for a particular purpose and noninfringement. In no event shall the authors or copyright holders be liable for any claim, damages or other liability, whether in an action of contract, tort or otherwise, arising from, out of or in connection with the software or the use or other dealings in the software.

Index

315

9 780367 394547